NORTH-HOLLAND
MATHEMATICS STUDIES

Notas de Matemática
editor: Leopoldo Nachbin

Saks Spaces and Applications to Functional Analysis

J.B. COOPER

NORTH-HOLLAND

SAKS SPACES
AND APPLICATIONS TO FUNCTIONAL ANALYSIS

NORTH-HOLLAND
MATHEMATICS STUDIES 28

Notas de Matemática (64)

Editor: Leopoldo Nachbin

*Universidade Federal do Rio de Janeiro
and University of Rochester*

Saks Spaces and Applications to Functional Analysis

JAMES BELL COOPER

Johannes-Kepler Universität, Linz, Austria

1978

NORTH-HOLLAND PUBLISHING COMPANY - AMSTERDAM • NEW YORK • OXFORD

North-Holland ISBN: 0 444 85100 3

PUBLISHERS:
NORTH-HOLLAND PUBLISHING COMPANY
AMSTERDAM•NEW YORK•OXFORD

SOLE DISTRIBUTORS FOR THE U.S.A. AND CANADA:
ELSEVIER / NORTH HOLLAND, INC.
52 VANDERBILT AVENUE, NEW YORK, N.Y. 10017

Library of Congress Cataloging in Publication Data

Cooper, James Bell.
Saks spaces and applications to functional analysis.

(Notas de matemática ; 64) (North-Holland mathematics studies)
Bibliography: p.
Includes index.
1. Saks spaces. 2. Functional analysis.
I. Title. II. Series.
QA1.N86 no. 64 [QA322] 510'.8s [515'.73] 78-1921
ISBN 0-444-85100-3

PRINTED IN THE NETHERLANDS

PREFACE

This monograph is concerned with two streams in functional analysis - the theories of mixed topologies and of strict topologies. The first deals with mathematical objects consisting of a vector space together with a norm and a locally convex topology which are in some sense compatible. Their theory can be regarded as a generalisation of that of Banach spaces, complementary to the theory of locally convex spaces. Although closely related to locally convex spaces, the theory of mixed spaces (or Saks spaces as we propose to call them) has its own flavour and requires its own special techniques. The second theory is concerned with a number of special topologies on spaces of functions and operators which possess the common property that they are substitutes for natural norm topologies which are not suitable for certain applications. The best known example of the latter is the strict topology on the space of bounded, continuous, complex-valued functions on a locally compact space, which was introduced by Buck. The connection between these two topics is provided by the fact that the important strict topologies can be regarded in a natural way as Saks spaces and this fact allows a simple and unified approach to their theory.

The author feels that the theory of Saks spaces is sufficiently well-developed and useful to justify an attempt at a first synthesis of the theory and its applications. The present monograph is the consequence of this conviction.

The book is divided into five chapters, devoted successively to the general theory of Saks spaces and to important spaces of functions or operators with strict topologies (spaces of bounded continuous functions, bounded measurable functions, operators on Hilbert spaces and bounded holomorphic functions respectively). An appendix contains a category-theoretical approach to Saks spaces. Each chapter is divided into sections and Propositions, Definitions etc. are numbered accordingly. Hence a reference to I.1.1 is to the first definition (for example) in § 1 of Chapter I. Within Chapter I this would be abbreviated to 1.1. At the end of the book, indexes of notations and terms have been added.

An attempt has been made to make the first five chapters as self-contained as possible so that they will be useful and accessible to non-specialists. However, some results have been given without proof, either because they are standard or because they seemed to the author to provide useful insight on the subject but the proofs were unsuitable for inclusion. Some of the latter are relegated to Remarks. In the first Chapter, only familiarity with the basic concepts of Banach spaces and locally convex spaces is assumed. In Chapter II - V the principle of GILLMAN and JERISON has been applied - namely to make them comprehensible to a reader who understands the words in their respective titles. The monograph should therefore be suitable say for a graduate course or seminar. However, by providing each Chapter with notes (brief historical remarks and references to related research articles) and a list of references, an attempt has been made to produce a work which could also be useful to research workers in this and related fields. In addition to those references which we have used directly in the preparation of this book, we have tried to include a complete bibliography of papers which relate to Saks spaces and their applications, including some where the connection may seem rather remote.

As a warning to the reader, we mention that all topological spaces (and hence also locally convex spaces, uniform spaces etc.) are tacitly assumed to be Hausdorff.

It goes without saying that the author gratefully acknowledges the contributions to this book of all those mathematicians whose published works, preprints, correspondence or conversation with the author have been used in its preparation. Special thanks are due to Fräulein G. Jahn who made a beautiful job of typing a manuscript in a foreign language under rather trying circumstances.

J.B. Cooper

Edramsberg, October, 1977

LIST OF CONTENTS

CHAPTER I - MIXED TOPOLOGIES

Introduction: The objects of study in this chapter are Banach spaces with a supplementary structure in the form of an additional locally convex topology. The motivation lies in the interplay between certain mathematical objects (topological spaces, measure spaces etc.) and suitable spaces of (complex-valued) functions on them. These often have a natural Banach space structure. However, by passing over from the original spaces to the associated Banach spaces, one frequently loses crucial information on the underlying space. A good example (which will be the subject of our most important application of mixed topologies) is the Banach space $C^{\infty}(S)$ of bounded, continuous, complex-valued functions on a locally compact space S where it is impossible to recover S from the Banach space structure of $C^{\infty}(S)$ (in contrast to the case of compact spaces S).
As we shall see in Chapter II, this situation can be saved by enriching the structure of $C^{\infty}(S)$ with the topology τ_K of uniform convergence on the compact subsets of S. The class of spaces that we consider can be regarded as a generalisation of the class of Banach spaces (we can "enrich" a Banach space in a trivial way, namely by adding its own topology). In fact these spaces can be regarded as projective limits of certain spectra of Banach spaces with contractive linking mappings (just as one can regard (complete) locally convex spaces as projective limits of arbitrary spectra of Banach spaces) and we shall lay particular emphasis on this fact for two reasons: for

purely technical reasons and secondly because, in applications to function spaces, we shall constantly use the fact that our function spaces are constructed out of simpler blocks which correspond exactly to the members of a representing spectrum of Banach spaces. As an example, dual to the fact that one can consider a locally compact space as being built up from its compact subspaces, we find that one can construct the space $C^{\infty}(S)$ from the spectrum defined by the spaces $\{C(K)\}$ as K runs through these subsets.

One of our main tools in the study of our enriched Banach spaces will be a natural locally convex topology - the mixed topology of the title of this chapter. It turns out that this can be regarded as a generalised type of inductive limit. The latter were systematically studied by GARLING in his dissertation. For this reason, we begin with a treatment of this theory in the generality suitable to our purposes.

For the convenience of the reader, we now give a brief summary of Chapter I. In the first section, we give a basic treatment of generalised inductive limits. Essentially, we consider a vector space with two locally convex topologies which satisfy suitable compatibility conditions. We then introduce in a natural way a "mixed topology" and this section is devoted to relating its properties to those of the original topologies. However, a closer examination of the definitions and results shows that, for one of the topologies, only the bounded sets are relevant. We have taken the consequences of this observation by replacing this topology by a "bornology", that is a suitable collection of sets which satisfy properties which one would expect of a family of bounded sets. We really only use the language (and not the theory) of bornologies and introduce explicitly all the terms that we use. In section 2, we give a list of examples of spaces with mixed topologies. Some of these will be studied in detail (and in more generality) in the following chapters. Others are introduced to supply counter-examples. All are used to illustrate the ideas of the first section. In section three, we define the class of enriched Banach spaces mentioned, restate the results of sections 1 in the form that we shall require them for applications and describe the usual methods for constructing new spaces (subspaces, products, tensor products etc.). It is perhaps not inappropriate to mention here that one of the main reasons for our emphasis on spaces with two structures (a norm and a locally convex topology) rather than on locally convex spaces of a rather curious type is the fact that it is important that these constructions be so carried out that this double structure is preserved and not in the sense of locally convex spaces. The fourth section is devoted to attempts to extend the classical results on Banach spaces to enriched Banach spaces (e.g. Banach-Steinhaus theorem, closed graph theorem). The results obtained are perhaps rather unsatisfactory since they involve special hypotheses but, as we shall see later, they can often be applied to important function spaces. In any case, there are simple counter-examples which show that such results cannot hold without rather special restrictions.

I.1. BASIC THEORY

As announced in the Introduction to this chapter, it is convenient for us to use the language of bornologies. We begin with their definition:

1.1. Definition: Let E be a vector space. A ball in E is an absolutely convex subset of E which does not contain a non-trivial subspace. If B is a ball in E, we write E_B for the linear span $\bigcup_{n=1}^{\infty} nB$ of B in E. Then

$$\| \ \|_B : x \longrightarrow \inf \{\lambda > 0 : x \in \lambda B\}$$

is a norm on E. If $(E_B, \| \ \|_B)$ is a Banach space, B is a Banach ball.

Note that any absolutely convex, bounded subset of a locally convex space is a ball. The following Lemma gives a sufficient (but not necessary) condition for it to be a Banach ball.

1.2. Lemma: Let B be a bounded ball in a locally convex space (E,τ). Then if B is sequentially complete for τ (and in particular if it is τ-complete), B is a Banach ball.

Proof: Let (x_n) be a Cauchy sequence in $(E_B, \| \|_B)$. Then, since B is bounded, (x_n) is τ-Cauchy. Hence there is an $x \in B$ so that $x_n \to x$ for τ. We show that $\|x_n - x\|_B \longrightarrow 0$.
If $\varepsilon > 0$, there is an $N \in \mathbb{N}$ so that $(x_m - x_n)$ belongs to εB for $m,n \geq N$. Since B (and so also εB) is sequentially complete and so sequentially closed, we can take the limit over n to deduce that $x_m - x$ belongs to εB for $m \geq N$.

1.3. Definition: If E is a vector space, a (convex) bornology on E is a family $\mathcal{B}$ of balls in E so that

(a) $E = \cup \mathcal{B}$;

(b) if $B \in \mathcal{B}$, $\lambda > 0$, then $\lambda B \in \mathcal{B}$;

(c) $\mathcal{B}$ is directed on the right by inclusion (i.e. if $B, C \in \mathcal{B}$ then there exists $D \in \mathcal{B}$ with $B \cup C \subseteq D$);

(d) if $B \in \mathcal{B}$ and C is a ball contained in B, then $C \in \mathcal{B}$.

A subset B of E is $\mathcal{B}$-bounded if it is contained in some ball in $\mathcal{B}$.
A basis for $\mathcal{B}$ is a subfamily $\mathcal{B}_1$ of $\mathcal{B}$ so that each $B \in \mathcal{B}$ is a subset of some $B_1 \in \mathcal{B}_1$.
$(E,\mathcal{B})$ is complete if $\mathcal{B}$ has a basis consisting of Banach balls.
$\mathcal{B}$ is of countable type if $\mathcal{B}$ has a countable basis.
If (E,τ) is a locally convex space, then $\mathcal{B}_\tau$, the family of all τ-bounded, absolutely convex subsets of E is a bornology on E - the von Neumann bornology. In many of our applications $\mathcal{B}$ will be the von Neumann bornology of a normed space $(E, \| \|)$. This is of countable type (the family $\{nB_{\| \|}\}_{n \in \mathbb{N}}$ where $B_{\| \|}$ is the unit ball of E is a basis).

Now we consider a vector space E with a locally convex topology τ and a bornology $\mathcal{B}$ of countable type which are compatible in the following sense:

$\mathcal{B} \subseteq \mathcal{B}_\tau$ and $\mathcal{B}$ has a basis of τ-closed sets. Then we can choose a basis (B_n) for $\mathcal{B}$ with the following properties:

(a) $B_n + B_n \subseteq B_{n+1}$ for each n;

(b) each B_n is τ-closed;

(if (C_n) is a countable basis for $\mathcal{B}$, we can define (B_n) inductively as follows: take $B_1 = C_1$. Once $B_1,\ldots,B_n$ have been chosen, we can find a τ-closed ball in $\mathcal{B}$ which contains $B_n + B_n + C_{n+1}$. This is our B_{n+1}). In future, we shall tacitly assume that a given basis (B_n) has the above properties.

1.4. <u>Definition</u>: We define the mixed locally convex structure $\gamma = \gamma[\mathcal{B},\tau]$ as follows:

Let $\mathcal{U} = (U_n)_{n=1}^{\infty}$ be a sequence of absolutely convex τ-neighbourhoods of zero and write

$$\gamma(\mathcal{U}) := \bigcup_{n=1}^{\infty} (U_1 \cap B_1 + \ldots\ldots + U_n \cap B_n).$$

Then the set of all such $\gamma(\mathcal{U})$ forms a base of neighbourhoods of zero for a locally convex structure on E and we denote it by $\gamma[\mathcal{B},\tau]$ (or simply by γ if no confusion is possible).

In the case that $\mathcal{B}$ is the bornology defined by a norm on E, we write $\gamma[\|\ \|,\tau]$ for the structure $\gamma[\mathcal{B},\tau]$.

The following Proposition gives a natural characterisation of γ.

1.5. <u>Proposition</u>: (i) γ is finer than τ;

(ii) γ and τ coincide on the sets of $\mathcal{B}$;

(iii) γ is the finest linear topology on E which coincides with τ on the sets of $\mathcal{B}$.

<u>Proof</u>: (i) if U is a τ-neighbourhood of zero, then $U \supseteq \gamma(U_n)$ where $U_n := 2^{-n}U$.

(ii) if $B \in \mathcal{B}$, we can choose a positive integer r so that $B - B \subseteq B_r$. A typical neighbourhood of the point $x_0 \in B$ for the topology induced by γ on B has the form

$$B \cap (x_0 + \gamma(U_n)).$$

Then $U_r \cap (B-B) \subseteq U_r \cap B_r \subseteq \gamma((U_n))$ and so

$$(x_0 + U_r) \cap B \subseteq (x_0 + \gamma((U_n)) \cap B.$$

(iii) let τ_1 be a linear topology on E which coincides with τ on the sets of $\mathcal{B}$. We show that γ is finer than τ_1. Let W be a neighbourhood of zero for τ_1 and choose neighbourhoods W_n of zero so that $W_0 = W$ and $W_n + W_n \subseteq W_{n-1}$ $(n \geq 1)$. There are τ-neighbourhoods (U_n) of zero so that $U_n \cap B_n \subseteq W_n$. Then, for any n

$$(U_1 \cap B_1) + \ldots\ldots + (U_n \cap B_n) \subseteq W$$

and so $\gamma((U_n)) \subseteq W$.

1.6. <u>Corollary</u>: (i) γ is independent of the choice of basis (B_n);

(ii) if τ and τ_1 are suitable locally convex topologies on E (i.e. if τ and τ_1 are compatible with $\mathcal{B}$) then $\gamma[\mathcal{B},\tau] = \gamma[\mathcal{B},\tau_1]$ if and only if τ and τ_1 coincide on the sets of $\mathcal{B}$.

The localisation property of γ expressed in I.1.5 implies that the continuity of linear mappings is determined by their behaviour on the bounded sets of E.

1.7. Corollary: Let H be a family of linear mappings from E into a topological vector space F. Then H is γ-equicontinuous if and only if $H|_B$ is τ-equicontinuous for each $B \in \mathcal{B}$. In particular, a linear mapping T from E into F is continuous if and only if $T|_B$ is τ-continuous for each $B \in \mathcal{B}$.

Proof: We must show that if W is a neighbourhood of zero in F then $\bigcap_{T \in H} T^{-1}(W)$ is a γ-neighbourhood of zero in E. But $\bigcap_{T \in H} (T|_B)^{-1}(W) = \bigcap_{T \in H} T^{-1}(W) \cap B$ and so we can apply 1.5.(iii).

In view of the above property, the following Lemma of GROTHENDIECK will be useful:

1.8. Lemma: Let (E,τ) be a locally convex space, B an absolutely convex subset of E and T a linear mapping from B into a locally convex space F. Then $T|_B$ is τ_B-continuous if and only if it is τ_B-continuous at zero.

Proof: Let V be an absolutely convex neighbourhood of zero in F, x a point of B. We must find a neighbourhood U of zero in E so that $T((x + U) \cap B) \subseteq Tx + V$. We choose U (absolutely convex) so that $T(B \cap (U/2)) \subseteq V/2$. Then if $y \in ((x+U) \cap B)$, $x-y \in B - B = 2B$ and the result follows from the inclusion $T(2B \cap U) \subseteq V$.

As we shall see later, the topology γ is, in the interesting cases, never metrisable (or even bornological). However, it does, sometimes, have one useful property in common with such spaces.

1.9. Proposition: Suppose that $\mathcal{B}$ has a basis of τ-metrisable sets. Then a linear mapping from E into a topological vector space F is continuous if and only if it is sequentially continuous.

Proof: For any $B \in \mathcal{B}$, $T|_B$ is sequentially continuous and so continuous. The result then follows from 1.7.

In the following Propositions, we characterise certain properties (boundedness, compactness, convergence) with respect to γ directly in terms of $\mathcal{B}$ and τ.

1.10. Proposition: A sequence (x_n) in E converges to x in (E,γ) if and only if $\{x_n\}$ is $\mathcal{B}$-bounded and $x_n \to x$ in (E,τ).

Proof: We can suppose, that $x = 0$. By 1.5 it suffices to show that if $x_n \to 0$ in (E,τ), then $\{x_n\}$ is $\mathcal{B}$-bounded. If this were false, we could find a subsequence (x_{n_k}) so that $x_{n_k} \notin B_k$.

Since B_k is τ-closed, we can choose a τ-neighbourhood of zero U_k so that $x_{n_k} \notin B_k + 2U_k$ and we can suppose that $U_k + U_k \subseteq U_{k-1}$ $(k > 1)$. Then for each $k > 1$

$$\begin{aligned} \gamma((U_n)) &= \bigcup_{n=1}^{\infty} (U_1 \cap B_1 + \ldots\ldots + U_n \cap B_n) \\ &\subseteq \bigcup_{p=1}^{\infty} (B_1 + \ldots\ldots + B_{k-1} + U_k + \ldots\ldots + U_{k+p}) \\ &\subseteq B_k + 2U_k . \end{aligned}$$

Hence $x_{n_k} \notin \gamma((U_n))$ for each k which contradicts the fact that (x_n) is a γ-null-sequence.

1.11. <u>Proposition</u>: A subset B of E is γ-bounded if and only if it is $\mathcal{B}$-bounded.

<u>Proof</u>: Suppose that B is $\mathcal{B}$-bounded. Then $B \subseteq B_r$ for some r. Let (U_n) be a sequence of absolutely convex neighbourhood of zero. Then there is a $K > 1$ so that $B \subseteq KU_r$. Hence

$$B \subseteq K(U_r \cap B_r) \subseteq K\gamma\,((U_n))$$

and so B is γ-bounded.

Now suppose that B is γ-bounded. If B were not $\mathcal{B}$-bounded, we could find a sequence (x_n) in B so that $x_n \notin nB_n$ for each positive integer n. Now $n^{-1}x_n \to 0$ in (E,γ) and so $\{n^{-1}x_n\}$ is $\mathcal{B}$-bounded - contradiction.

1.12. <u>Proposition</u>: A subset A of E is γ-compact (precompact, relatively compact) if and only if it is $\mathcal{B}$-bounded and τ-compact (precompact, relatively compact).

Proof: This follows immediately from 1.11 and 1.5.(ii).

We recall that a locally convex space is semi-Montel if its bounded sets are relatively compact. It is Montel if, in addition, it is barrelled. In the next Proposition, we characterise semi-Montel mixed spaces. As we shall see below, non-trivial mixed topologies are never barrelled - and so never Montel.

1.13. Proposition: (E,γ) is semi-Montel if and only if $\mathcal{B}$ has a basis of τ-compact sets.

Proof: This is a direct consequence of 1.12 and the definition.

We now consider the completeness of (E,γ). It is customary to use here a completeness theorem of RAIKOV which generalises KÖTHE'S completeness theorem. However, this result is rather inaccessible so we give a direct proof, using a method of DE WILDE and HOUET.

1.14. Proposition: (E,γ) is complete if and only if $\mathcal{B}$ has a basis of τ-complete sets.

Proof: The necessity of the given condition follows from 1.5.(ii) and 1.11.
Now suppose that we have a basis (B_n) for $\mathcal{B}$ consisting of τ-complete sets. Let $(\hat{E},\hat{\gamma})$ be the locally convex completion of E. If $E \neq \hat{E}$, choose $x_o \notin \hat{E} \setminus E$. Then for each n, $x_o \notin 2B_n$

and so by the Hahn-Banach theorem ([56], p.39, Folg. 2), there is an $f_n \varepsilon (\hat{E})'$ with $f_n \varepsilon B_n^o$ and $f_n(x_o) = 2$. Since only finitely many f_n are non-zero on a given B_m, $\{f_n\}$ is equicontinuous on E (1.7) and hence also on $\hat{E}$. Then by the theorem of ALAOGLU-BOURBAKI ([31], p.250, § 10.10(4)) $\{f_n\}$ is $\sigma(E',\hat{E})$ compact and so has a $\sigma(E',\hat{E})$-limit point f. Then f vanishes on each B_m and so on E. Thus f = 0 which contradicts the fact that $f(x_o) = 2$.

We now make some remarks on the case where $\mathcal{B}$ is the bornology associated with a norm on E. Then there are three natural locally convex topologies τ,γ and $\tau_{\|\ \|}$, the norm topology, on E and we discuss their distinctness. We first note that the equality $\tau = \gamma$ means essentially that τ is already a mixed topology i.e. we have gained nothing by mixing. On the other hand the condition $\gamma = \tau_{\|\ \|}$ means that we are in the trivial situation (trivial from the point of view of mixed topologies) of a normed space "enriched" by its own topology (since $\gamma[\|\ \|,\tau] = \gamma[\|\ \|,\tau_{\|\ \|}]$). The following result shows that if γ belongs to the traditional classes of well-behaved locally convex spaces, then we have this trivial situation.

1.15. Proposition: If γ is bornological (in particular, metrisable) or barrelled, then $\tau = \tau_{\|\ \|}$.

Proof: If γ is bornological, then the identity mapping from (E,γ) into $(E,\|\ \|)$, being bounded (1.11) is continuous

and so $\gamma \supseteq \tau_{\|\ \|}$. The inverse inequality is obvious.
If γ is barrelled, then $B_{\|\ \|}$, the unit ball of $(E, \|\ \|)$, being a barrel in (E,γ), is a γ-neighbourhood of zero (we are assuming that $B_{\|\ \|}$ is τ-closed - strictly speaking, this need not to be the case. However, it follows easily from the compatibility conditions that we can find an equivalent norm so that this condition is satisfied).

The essential property of γ is that given in 1.5.(iii). The neighbourhood basis used in the definition was chosen so that this would hold. However, in applications, we shall frequently require a much less obvious description of γ-neighbourhoods of zero. Let $\mathcal{U}$ be as in 1.4 (except that it is now convenient to index from zero to infinity) and put

$$\tilde{\gamma}(\mathcal{U}) := U_0 \cap \bigcap_{n=1}^{\infty} (U_n + B_n).$$

Then the family of such sets forms a neighbourhood basis for a locally convex topology on E which we denote by $\tilde{\gamma}[\mathcal{B},\tau]$.

1.16. <u>Proposition</u>: $\tilde{\gamma}[\mathcal{B},\tau] = \gamma[\mathcal{B},\tau]$.

<u>Proof</u>: Firstly, $\tilde{\gamma}$ is coarser than τ on each set B_k.
For $\tilde{\gamma}(\mathcal{U}) \cap B_k = (U_0 \cap \bigcap_{n=1}^{k} (B_n + U_n)) \cap B_k$ and this is a $\tau|_{B_k}$-neighbourhood of zero. Hence by 1.5.(iii), γ is finer than $\tilde{\gamma}$.
Now we show that γ is coarser than $\tilde{\gamma}$. Let $\gamma((U_n))$ be a typical γ-neighbourhood of zero. There exists a decreasing sequence

$(V_n)_{n=0}^{\infty}$ of τ-neighbourhoods of zero so that $V_{n-1} + V_{n-1} \subseteq U_{n+1}$ $(n = 0,1,2,\ldots)$. We shall show that $\tilde{\gamma}((V_n)) \subseteq \gamma((U_n))$. Choose $x \in \tilde{\gamma}((V_n))$. Then $x \in V_0$ and, for each n, x has a decomposition $y_n + z_n$ where $y_n \in B_n$, $z_n \in V_n$. Define $x_1 := y_1$, $x_n := y_n - y_{n-1}$ $(n > 1)$. Then

$$x_1 + \ldots + x_n + z_n = y_1 + (y_2 - y_1) + \ldots + (y_n - y_{n-1}) + z_n = y_n + z_n = x$$

and $z_{n-1} = x_n + z_n$.

We have $x_n = z_{n-1} - z_n \in V_{n-1} + V_n \subseteq V_{n-1} + V_{n-1} \subseteq U_{n+1}$

and $x_n = y_n - y_{n-1} \in B_n + B_{n-1} \subseteq B_{n+1}$.

Hence $x_n \in U_{n+1} \cap B_{n+1}$.

If n_0 is so chosen that $x \in B_{n_0}$, then $z_{n_0} = x - y_{n_0} \in B_{n_0+1}$.

On the other hand, $z_{n_0} \in V_{n_0} \subseteq U_{n_0+2}$.

Hence we have

$$x = x_1 + \ldots + x_n + z_n \in U_2 \cap B_2 + \ldots + U_{n_0+1} \cap B_{n_0+1} + U_{n_0+2} \cap B_{n_0+2}$$

$$\subseteq \gamma((U_n)).$$

1.17. <u>Corollary</u>: γ has a basis consisting of τ-closed sets.

<u>Proof</u>: If U_n is an absolutely convex τ-neighbourhood of zero, then

$$(2^{-1}U_n) + B_n \subseteq \overline{(2^{-1}U_n) + B_n} \subseteq U_n + B_n$$

and so $\tilde{\gamma}((2^{-1}U_n)) \subseteq U_0 \cap \bigcap_{n=1}^{\infty} \overline{((2^{-1}U_n) + B_n)} \subseteq \tilde{\gamma}((U_n))$

and this implies the result.

We now consider duality for (E,γ). E has three dual spaces:

E'_τ - the dual of the locally convex space (E,τ);

E'_γ - the dual of the locally convex space (E,γ);

$E'_\mathcal{B}$ - the dual of the space $(E,\mathcal{B})$, that is, the space of linear forms on E which are <u>bounded</u> i.e. bounded on the sets of $\mathcal{B}$.

Then $E'_\tau \subseteq E'_\gamma \subseteq E'_\mathcal{B}$ and we regard each of these spaces as a locally convex space with the topology of uniform convergence on the τ-bounded sets (resp. the γ-bounded sets, resp. the sets of $\mathcal{B}$). Since $\mathcal{B}$ is of countable type, E' is metrisable and it is also clearly complete (since the uniform limit of bounded functions is bounded). Hence it is a Fréchet space. Our next results characterise E'_γ and its equicontinuous subsets in terms of E'_τ and $E'_\mathcal{B}$.

1.18. <u>Proposition</u>: (i) E'_γ is a locally convex subspace of $E'_\mathcal{B}$; (ii) E'_γ is the closure of E'_τ in $E'_\mathcal{B}$ and so is a Fréchet space.

<u>Proof</u>: (i) follows directly from 1.11.
(ii) E'_γ is closed in $E'_\mathcal{B}$ since the limit of a sequence in E'_γ is continuous on the sets of $\mathcal{B}$ and so is in E'_γ by 1.7.
We show that E'_τ is dense in E'_γ. Let B be a τ-closed ball in $\mathcal{B}$ and ε be a positive number. If $f \in E'_\gamma$ then there is an absolutely convex τ-neighbourhood U of zero so that $|f(x)| \le \varepsilon$ if $x \in B \cap U$ i.e. f belongs to $\varepsilon(B \cap U)^{\circ}$ (polar in E^*, the algebraic dual of E). Now the polar of $B \cap U$ is the closure

of $1/2(B^o + U^o)$ in $\sigma(E^*,E)$. But this set is closed since U^o is $\sigma(E^*,E)$-compact by the theorem of ALAOGLU-BOURBAKI and so

$$(B \cap U)^o \subseteq B^o + U^o.$$

Hence f belongs to $\varepsilon(U^o + B^o)$ and so there is a g belonging to $\varepsilon U^o \subseteq E'_\tau$ such that f-g belongs to εB^o i.e. $|f(x)-g(x)| < \varepsilon$ if $x \in B$.

1.19. Corollary: Let τ_1, τ_2 be locally convex topologies on E which are compatible with $\mathcal{B}$ and suppose that τ_1 and τ_2 have the same dual. Then $\gamma[\mathcal{B},\tau_1]$ and $\gamma[\mathcal{B},\tau_2]$ have the same dual.

1.20. Proposition: A subset B of E is γ-weakly compact if and only if it is $\mathcal{B}$-bounded and $\sigma(E,E'_\tau)$-compact.

Proof: The condition is clearly necessary. It is sufficient since if B is $\mathcal{B}$-bounded then, regarded as a subset of the dual of E'_γ, it is equicontinuous and so the weak topologies defined by E'_γ and its dense subspaces E'_τ coincide on it (see [61], p.83, III.4.5).

1.21. Corollary: (E,γ) is semi-reflexive if and only if $\mathcal{B}$ has a basis of $\sigma(E,E'_\tau)$-compact sets.

1.22. Proposition: A subset H of $E'_{\mathcal{B}}$ is γ-equicontinuous if and only if it satisfies the following condition:

for every strong neighbourhood U of zero in E'_β, there is a τ-equicontinuous set H_1 in E'_τ so that

$$H \subseteq U + H_1.$$

Proof: Sufficiency: choose $B \in \mathcal{B}$, $\varepsilon > 0$. It is sufficient to find a τ-neighbourhood V of zero so that if $x \in B \cap V$, $f \in H$, then $|f(x)| \leq \varepsilon$. We choose V so that

$$H \subseteq (\varepsilon/2)B^o + (\varepsilon/2)V^o \subseteq \varepsilon(V \cap B)^o$$

(for the last inclusion, cf. the proof of 1.18.(ii)).

Necessity: suppose that H is γ-equicontinuous and U is a strong neighbourhood of zero in E'_γ. We can suppose that $U = B_k^o$ for some positive integer k. Then there is a γ-neighbourhood of zero $\gamma((U_n))$ so that

$$H \subseteq \{\gamma((U_n))\}^o \subseteq (U_1 \cap B_1 + \ldots + U_k \cap B_k)^o$$
$$\subseteq (U_k \cap B_k)^o \subseteq U_k^o + B_k^o.$$

1.23. Corollary: Let (x_n) be a null-sequence in E'_γ. Then $\{x_n\}$ is γ-equicontinuous.

Proof: If U is a strong neighbourhood of zero in E'_β, then all but finitely many of the elements of the sequence lie in U. Hence we can apply 1.22.

1.24. Corollary: Let A be a compact subset of E'_γ. Then A is γ-equicontinuous.

Proof: Since E'_γ is a Fréchet space, A is contained in the closed, absolutely convex hull of a null sequence ([56], Ch.7, § 2, Lemma 2). By 1.23 the range of this sequence is equicontinuous and hence so is its $\sigma(E'_\gamma,E)$-closed absolutely convex hull and this set contains A.

If (F,τ_1) is a locally convex space and E is a subspace, there is a natural vector space isomorphism between F'/E^o and E' induced by the restriction mapping from F' onto E'. Since the latter mapping is continuous for the strong topologies, the map from F'/E^o onto E' is continuous. In general, it is not a locally convex isomorphism. This is the case, however, when E has a mixed topology.

1.25. Proposition: Let (E,γ) be a locally convex subspace of (F,τ_1). Then the strong topology on F'/E^o as the dual of (E,γ) coincides with the quotient of the strong topology on F'.

Proof: We show that the natural mapping from E'_γ onto F'/E^o is continuous. Since E'_γ is a Fréchet space, it suffices to show that every sequence which converges to zero in E'_γ is bounded in F'/E^o. But such a sequence is γ-equicontinuous (1.23) and so is the restriction of an equicontinuous set in F' (Hahn-Banach theorem). The image of such a set in F'/E^o is bounded.

1.26. Corollary: If, in addition, E is dense in F, then every bounded set in F is contained in the closure of a bounded set in E.

Proof: By 1.25, the natural vector space isomorphism between the duals of E and F is an isomorphism for the strong topologies and this is equivalent to the statement of the Corollary (bipolar theorem).

1.26 can be used to give an alternative proof of 1.14.

1.27. Remark: It follows from the results of this paragraph that the spaces of type $(E,\gamma[\mathcal{B},\tau])$ can be internally characterised as the class of locally convex spaces (E,τ) with the following properties:

a) $\mathcal{B}_\tau$ is of countable type;

b) an absolutely convex subset V of E is a neighbourhood of zero if and only if $V \cap B$ is a τ-neighbourhood of zero for each $B \in \mathcal{B}_\tau$.

Hence they form a generalisation of the class of (DF)-spaces of GROTHENDIECK. They have been studied from this point of view by NOUREDDINE who calls then D_b-spaces (see [41]). NOUREDDINE has shown that they possess many of the properties of (DF)-spaces. We mention without proof the following:

I. if (V_n) is a sequence of neighbourhoods of zero in E, there is a sequence (α_n) of positive scalars so that $\bigcap \alpha_n V_n$ is a neighbourhood of zero.

II. every continuous linear mapping from E into a metrisable locally convex space F is bounded (i.e. it carries some neighbourhood of zero in E into a bounded set in F).

III. the class of D_b-spaces is closed under the formation of quotients, countable limits, subspaces of finite codimension and even subspaces of countable co-dimension when the space is complete.

IV. a D_b-space E has the property (B) of PIETSCH and so every absolutely summable sequence is totally summable (i.e. is absolutely summable in the normed space $(E_B, \| \|_B)$ for some $B \in \mathcal{B}_\tau$). As a consequence, the theorem of DVORETSKY-ROGERS holds in a simple D_b-space (i.e. a D_b-space whose von Neumann bornology is generated by a norm): if E is infinite dimensional, then E possesses a sequence which is summable but not absolutely summable.

I.2. EXAMPLES

A. Let F be a Fréchet space and denote its dual by E. On E we consider the structures:

$\mathcal{B}$:= the collection of absolutely convex, equicontinuous subsets of E (equivalently, the bounded sets of E, provided with the strong topology as the dual of F);

$\sigma(E,F)$ - the weak topology defined on E by F.

Then the triple $(E, \mathcal{B}, \sigma(E,F))$ satisfies the conditions of 1.4 and so we can form the associated mixed topology. We identify this topology. Firstly, the dual of $(E, \sigma(E,F))$ is, of course, F and on F the topology of uniform convergence on $\mathcal{B}$ is the original topology. Hence, by 1.17, the dual of

(E,γ) is F. Now the $\sigma(E,F)$-equicontinuous subsets of F are the bounded, finite dimensional subsets. Hence, by 1.22 and a standard characterisation of precompact subsets of a locally convex space ([56], p.58), the γ-equicontinuous subsets of F are precisely the precompact subsets i.e. γ is the topology $\tau_c(E,F)$ of uniform convergence on the compact subsets of F.

B. If S is a set and $p \in [1,\infty]$, $\ell^p(S)$ denotes the space of complex functions x on S so that $\|x\|_p < \infty$ where

$$\|x\|_p := \Big(\sum_{t\in S} |x(t)|^p\Big)^{1/p} \qquad (p \in [1,\infty[)$$

$$\|x\|_\infty := \sup\{|x(t)| : t \in S\}.$$

$(\ell^p(S),\|\ \|_p)$ is a Banach space and we consider the topology τ_s of pointwise convergence on S. Then $(\ell^p(S),\|\ \|_p,\tau_s)$ satisfies the conditions of 1.4 and we can define the corresponding mixed topology. For $p > 1$, $\ell^p(S)$ is the dual of $\ell^q(S)$ where $1/p + 1/q = 1$ $(p > 1)$ and $q = 1$ $(p = \infty)$. Hence, since on the unit ball of $\ell^p(S)$ τ_s coincides with $\sigma(\ell^p(S),\ell^q(S))$ (compactness !) we are in the situation of A and so $\gamma = \tau_c(\ell^p,\ell^q)$. Similarly, for $p = 1$, $\ell^1(S)$ is the dual of

$$c_o(S) := \{x \in \ell^\infty(S) : \text{for each } \varepsilon > 0, \text{ there is a finite subset J of S so that } |x(t)| < \varepsilon \text{ for } t \notin J\}$$

and so γ is $\tau_c(\ell^1,c_o)$.

C. If $p < p_1$, then $\ell^p(S) \subseteq \ell^{p_1}(S)$ and so we can consider on $\ell^p(S)$ the mixed structure $\gamma[\|\ \|_p, \tau_{p_1}]$ where τ_{p_1} denotes the topology induced on $\ell^p(S)$ by $\|\ \|_{p_1}$. This mixed topology is distinct from that defined in B since the unit ball of $\ell^p(S)$ is not $\|\ \|_{p_1}$-compact. It follows from 1.18 that the dual of $\ell^p(S)$ under this mixed topology is $\ell^q(S)$ $(p > 1)$ and the dual of $\ell^1(S)$ is $c_o(S)$.

D. Let T be a locally compact space, S a collection of subsets of T so that $\bigcup S$ is dense in T. We denote by $C^\infty(T)$ the space of bounded, continuous functions from T into $\mathbb{C}$. Then

$$\|\ \|_\infty : x \longmapsto \sup\{|x(t)| : t \in T\}$$

is a norm on $C^\infty(T)$ and $(C^\infty(T), \|\ \|_\infty)$ is a Banach space. If $A \subseteq T$ then

$$p_A : x \longmapsto \{\sup |x(t)| : t \in A\}$$

is a seminorm on $C^\infty(T)$ and we denote by τ_S the locally convex structure generated by $\{p_A : A \in S\}$. Then $(C^\infty(T), \|\ \|_\infty, \tau_S)$ satisfies the conditions of 1.4 and so we can form the mixed topology $\gamma[\|\ \|_\infty, \tau_S]$ which we denote by β_S.

E. Let G be an open subset of the complex plane and denote by $H^\infty(G)$ the subspace of $C^\infty(G)$ consisting of holomorphic functions. Then $(H^\infty(G), \|\ \|_\infty, \tau_K)$ where K is the family of compact subsets of G and $\|\ \|_\infty$ and τ_K are defined as in D) satisfies the conditions of 1.4. We denote the corresponding mixed topology by β.

F. Let S be a locally compact space. $C_{oo}(S)$ denotes the set of functions in $C^{\infty}(S)$ with compact support. Thus $C_{oo}(S) = \bigcup_{K \varepsilon \mathcal{K}(S)} C_K(S)$ where $\mathcal{K}(S)$ denotes the family of compact subsets of S and $C_K(S)$ denotes the functions in $C^{\infty}(S)$ which have support in K. $C_K(S)$, with the norm induced from $C^{\infty}(S)$ is a Banach space. We define a bornology $\mathcal{B}$ on $C_{oo}(S)$ as follows: $B \varepsilon \mathcal{B}$ if and only if there is a $K \varepsilon \mathcal{K}(S)$ so that B is a bounded ball in $C_K(S)$. If S is σ-compact (i.e. the union of countably many compact subsets), then $\mathcal{B}$ is of countable type. Then $(C_{oo}(S), \mathcal{B}, \tau_{\mathcal{K}})$ satisfies the conditions of 1.4.

G. Let (F_n) be a sequence of Banach spaces and let

$$E := \sum_{n=1}^{\infty} F_n; \qquad F := \prod_{n=1}^{\infty} F_n$$

the vector space direct sum and product respectively. On F we consider the product topology τ and on E the von Neumann bornology of the direct sum locally convex topology on E (so that a set $B \subseteq E$ is bounded if it is bounded in a subspace of the form $\sum_{n=1}^{m} F_n$ for some m). Then the conditions of 1.4 are satisfied.

H. Let F be a Fréchet space, G a Banach space and denote by E the space L(F,G) of continuous linear mappings from F into G. If $\mathcal{B}$ is a bornology on F, contained in the von Neumann bornology, we define on E the following structures:

$\mathcal{B}_{eq}$ - the bornology generated by the equicontinuous balls in E;

$\tau_{\mathcal{B}}$ - the topology of uniform convergence on the sets of $\mathcal{B}$.

Then the conditions of 1.4 are satisfied.

I. Let E be a C^*-algebra with unit e, $S(E)$ the set of states of E i.e. the positive linear forms $f \in E'$ with $f(e) = 1$. Then each $f \in S(E)$ defines a seminorm

$$p_f : x \longmapsto \{f(x^*x)\}^{1/2}$$

on E. If $x \in E$

$$\|x\| = \sup \{p_f(x) : f \in S(E)\}$$

and so we can form the mixed topology $\gamma[\| \|, \tau_S]$ where τ_S is the locally convex topology generated by the family $\{p_f : f \in S(E)\}$.

J. Let (X,d) be a metric space. For convenience, we suppose that the associated topological space is compact. If $x \in C(X)$, x is <u>Lipschitz</u> if

$$\|x\|_L := \sup \{\frac{|x(s) - x(t)|}{d(s,t)} : (s,t) \in X \times X \setminus \Delta\}$$

(Δ the diagonal set) is finite. We denote by Lip (X) the space of such functions. On Lip (X) we consider the following norms :

$\| \|_\infty$ - the supremum norm;

$\| \|$ - the norm $x \longmapsto \max\{\|x\|_\infty, \|x\|_L\}$

Then $(\text{Lip}(X), \| \|, \tau_{\| \|_\infty})$ satisfies the conditions of 1.4.

K. Let $(E,\| \|)$ be a Banach space with basis (x_n) i.e. (x_n) is a sequence in E with the property that for every $x \in E$ there is a unique sequence (λ_n) of scalars so that $x = \sum_{n=1}^{\infty} \lambda_n x_n$. Then it is classical that

$$|||x||| := \sup\{\| \sum_{k=1}^{n} \lambda_k x_k\| : n \in \mathbb{N}\}$$

is a norm on E, equivalent to $\| \|$.
On E the seminorms

$$p_n : x \longmapsto \| \sum_{k=1}^{n} \lambda_k x_k \|$$

define a locally convex topology τ. Then $(E,||| \; |||,\tau)$ satisfies the conditions of 1.4.

1.3. SAKS SPACES

In this section we consider special types of spaces with mixed topologies - those whose bornology is induced by a norm. We propose to call them Saks spaces since they coincide essentially with the spaces introduced under this name by ORLICZ (the precise relationship between these concepts is discussed in the notes). We are concerned here with the basic constructions on Saks spaces. Since these are based on the corresponding constructions on Banach spaces (which they generalise), we recall the latter briefly (see SEMADENI [66]).
The construction of subspaces and quotient spaces of normed spaces is well-known. If $\{(E_\alpha,\| \|_\alpha)\}_{\alpha \in A}$ is a family of normed

spaces, we define new normed spaces as follows:

denote by E the Cartesian product $\prod_{\alpha\varepsilon A} E_\alpha$ and define extended norms

$$\| \|_1 : x = (x_\alpha) \longmapsto \sum_{\alpha\varepsilon A} \|x_\alpha\|_\alpha$$

$$\| \|_\infty : x = (x_\alpha) \longmapsto \sup_{\alpha\varepsilon A} \|x_\alpha\|_\alpha$$

on E. Then if

$$E_1 := \{x \varepsilon E : \|x\|_1 < \infty\}$$

$$E_\infty := \{x \varepsilon E : \|x\|_\infty < \infty\}$$

$(E_1, \| \|_1)$ and $(E_\infty, \| \|_\infty)$ are normed spaces. They are Banach if (and only if) each E_α is a Banach space. We write $B \sum_{\alpha\varepsilon A} E_\alpha$ and $B \prod_{\alpha\varepsilon A} E_\alpha$ for E_1 and E_∞ resp. They satisfy the universal property that one expects of a sum and a product if we restrict attention to linear contractions.

If A is a directed set and

$$\{\pi_{\beta\alpha} : E_\beta \longrightarrow E_\alpha, \alpha \leq \beta, \alpha,\beta \varepsilon A\}$$

$$(\text{resp. } \{i_{\alpha\beta} : E_\alpha \longrightarrow E_\beta, \alpha \leq \beta, \alpha,\beta \varepsilon A\})$$

is a projective spectrum (resp. an inductive spectrum) of normed spaces (i.e. each $\pi_{\beta\alpha}$ and each $i_{\alpha\beta}$ is a linear contraction with $\pi_{\alpha\alpha} = id_{E_\alpha}$ (resp. $i_{\alpha\alpha} = id_{E_\alpha}$) for each α and, if $\alpha \leq \beta \leq \gamma$, then $\pi_{\gamma\alpha} = \pi_{\beta\alpha} \circ \pi_{\gamma\beta}$ (resp. $i_{\alpha\gamma} = i_{\beta\gamma} \circ i_{\alpha\beta}$)), then we define the projective limit of the first spectrum as the (closed) subspace

$$\{(x_\alpha) \varepsilon B \prod_{\alpha\varepsilon A} E_\alpha : \pi_{\beta\alpha}(x_\beta) = x_\alpha \text{ for } \alpha \leq \beta\}$$

and denote it by $B\text{-}\varprojlim\{E_\alpha, \pi_{\beta\alpha}\}$. Similarly, the inductive limit of the second spectrum is the quotient of $B \sum_{\alpha\varepsilon A} E_\alpha$ with respect to the closed subspace generated by elements of the form $(x_\gamma - i_{\beta\gamma}(x_\beta))$ (we are regarding each space E_β as a subspace of $B\sum_{\alpha\in A} E_\alpha$ in the obvious way). In fact we shall only require the following special representation of an inductive limit: suppose that each E_α is a closed subspace of a given Banach space F and that A is so ordered that $\alpha \leq \beta$ if and only if $E_\alpha \subseteq E_\beta$ (and then $i_{\alpha\beta}$ is the natural injection): then the inductive limit is naturally identifiable with the closure of $\bigcup_{\alpha\in A} E_\alpha$ in F (see SEMADENI [66], § 11.8.3, p.212).

3.1. Lemma: Let (E,τ) be a locally convex space, $\|\ \|$ a norm on E with unit ball $B_{\|\ \|}$. Then the following are equivalent:

(a) $B_{\|\ \|}$ is τ-closed;

(b) $\|\ \|$ is lower semi-continuous for τ;

(c) $\|\ \| = \sup\{ p : p$ is a τ-continuous seminorm with $p \leq \|\ \|\}$.

Proof: (c) $\Longrightarrow$ (b) and (b) $\Longrightarrow$ (a) follow immediately from the elementary properties of semi-continuous functions.

(a) $\Longrightarrow$ (c) : suppose $x \in E$ with $\|x\| > 1$ i.e. $x \notin B_{\|\ \|}$. We need only find a continuous seminorm p on E so that $p \leq \|\ \|$ and $p(x) > 1$. By the Hahn-Banach theorem, there is an $f \in (E,\tau)'$ so that $\|f\| \leq 1$ on $B_{\|\ \|}$ and $f(x) > 1$. Then $\|f\|$ is such a seminorm.

3.2. Definition: A Saks space is a triple $(E,\|\ \|,\tau)$ where E is a vector space, τ is a locally convex topology on E and $\|\ \|$ is a norm on E so that $B_{\|\ \|}$, the unit ball of $(E,\|\ \|)$, is τ-bounded and satisfies one of the conditions of 3.1. If $(E,\|\ \|,\tau)$ and $(F,\|\ \|_1,\tau_1)$ are Saks spaces, a morphism from E into F is a linear norm contraction from E into F so that $T|B_{\|\ \|}$ is τ-τ_1-continuous. A Saks space $(E,\|\ \|,\tau)$ is complete if $B_{\|\ \|}$ is τ-complete. Then $(E,\|\ \|)$ is a Banach space (1.2).

In constructing Saks spaces, one occasionally produces triples $(E,\|\ \|,\tau)$ where all but the last condition (on the closure of $B_{\|\ \|}$) is satisfied. This forces us to take the following precaution: we define $B_1 := \overline{B}_{\|\ \|}$ (closure in τ). Then the Minkowski functional $\|\ \|_1$ of B_1 (as defined in 1.1) is a norm on E so that $(E,\|\ \|_1,\tau)$ is a Saks space. The following Lemma ensures that we do not lose any morphisms in this process :

3.3. Lemma: Let T be a linear mapping from E into a locally convex space F so that $T|_{B_{\|\ \|}}$ is τ-continuous. Then $T|_{B_1}$ is τ-continuous.

Proof: By 1.8, it suffices to show that $T|_{B_1}$ is continuous at zero. Let U be an absolutely convex neighbourhood of zero in F and choose an open neighbourhood V of zero in E so that $T(V \cap B_{\|\ \|}) \subseteq 1/2\ U$. Then $T(V \cap B_1) \subseteq U$. For if $x \in V \cap B_1$ and we choose $\lambda \in]0,1[$ so that $\lambda x \in B_{\|\ \|}$, then we can, by the continuity of T in $B_{\|\ \|}$, find $y \in V \cap B_{\|\ \|}$

so that $T(\lambda x) - T(\lambda y) \in \lambda/2\ U$. Then

$$Tx = \lambda^{-1}(T(\lambda x) - T(\lambda y)) + Ty \in U.$$

3.4. <u>The associated topology</u>: If $(E, \| \|, \tau)$ is a Saks space, we can form the mixed topology $\gamma[\| \|, \tau]$ - it is called the <u>associated locally convex topology</u> of E. Then a morphism between two Saks spaces is continuous for the associated topologies (1.5.(ii) and 1.7) and a Saks space is complete if and only if (E,γ) is a complete locally convex space (1.14). We repeat that, despite these facts (and others to follow), the relevant structure is that of a Saks space and not a locally convex space (this is one of the reasons that we have been careful not to forget the norms in the definition of a morphism - thus a γ-continuous linear mapping need not be a morphism although it is, of course, a scalar multiple of one). Hence we shall stubbornly persist in defining notions like completeness, compactness etc. in terms of the structure as a Saks space even when these can be expressed in terms of γ (using the theory of § I.1).

3.5. <u>Subspaces and quotient spaces</u>: Let $(E, \| \|, \tau)$ be a Saks space, F a vector subspace of E. Then if $\| \|_F, \tau_F$ denote the norm (resp. the locally convex topology) induced on F, $(F, \| \|_F, \tau_F)$ is a Saks space. We shall see later that $\gamma[\| \|_F, \tau_F]$ need not coincide with $\gamma[\| \|, \tau]|_F$.
If F is a γ-closed subspace, then we denote by $_F\| \|$ and $_F\tau$ the structures induces on the quotient space E/F. The triple

$(E/F, {}_F\|\ \|, {}_F\tau)$ need not be a Saks space since it can happen that the unit ball of $(E/F, {}_F\|\ \|)$ is not ${}_F\tau$-closed. However, by the process described before Lemma 3.3, we can obtain a Saks space which we shall call the quotient Saks space.

3.6. Completions: Let $(E,\|\ \|,\tau)$ be a Saks space and denote by $\hat{E}_\tau$ the completion of the locally convex space (E,τ). We write $\hat{B}$ for the closure of $B_{\|\ \|}$ in $\hat{E}_\tau$ and $\hat{E}$ for the linear span of $\hat{B}$ in $\hat{E}_\tau$. Then if $\|\ \|^\wedge$ is the Minkowski functional of $\hat{B}$ and $\hat{\tau}$ is the locally convex structure induced on $\hat{E}$ from $\hat{E}_\tau$, $(\hat{E},\|\ \|^\wedge,\hat{\tau})$ is a complete Saks space. We call it the (Saks space) completion of E. It has the following universal property: for every morphism T from $(E,\|\ \|,\tau)$ into a complete Saks space $(F,\|\ \|_1,\tau_1)$, there is a unique morphism $\hat{T}$ from $(\hat{E},\|\ \|^\wedge,\hat{\tau})$ into $(F,\|\ \|_1,\tau_1)$ which extends T (for we can extend T firstly to $\hat{B}$ by uniform continuity and then to $\hat{E}$ by linearity).

As an amusing example, consider the Saks space $(E,\|\ \|,\sigma(E,E'))$ where $(E,\|\ \|)$ is a normed space. The completion of $(E,\sigma(E,E'))$ is $(E')^*$ the algebraic dual of E' (ROBERTSON and ROBERTSON [56], Satz 18, p.71). The closure of B in $(E')^*$ is its bipolar, i.e. the unit ball of E", the bidual of $(E,\|\ \|)$. Hence the completion of $(E,\|\ \|,\sigma(E,E'))$ is E", with the Saks space structure described in 2.1 (as the dual of E'). Thus we can regard the bidual as a completion.

3.7. Products and projective limits: Let $\{(E_\alpha, \|\ \|_\alpha, \tau_\alpha)\}_{\alpha \in A}$ be a family of Saks spaces. We can give $(E_\infty, \|\ \|_\infty)$, the normed space product of $\{E_\alpha\}$, a Saks space structure by considering on E_∞ the topology τ_∞, the product of $\{\tau_\alpha\}$. Then the unit ball $B_{\|\ \|_\infty}$ of E_∞ is the product $\prod_{\alpha \in A} B_{\|\ \|_\alpha}$ and so is τ_∞-closed. For the same reason, $(E_\infty, \|\ \|_\infty, \tau_\infty)$ is complete if and only if each $(E_\alpha, \|\ \|_\infty, \tau_\infty)$ is. We call $(E_\infty, \|\ \|_\infty, \tau_\infty)$ the Saks space product of $\{E_\alpha\}$ and denote it by $S \prod_{\alpha \in A} E_\alpha$.

If $\{\pi_{\beta\alpha} : E_\beta \longrightarrow E_\alpha,\ \alpha \leq \beta,\ \alpha, \beta \in A\}$ is a projective system of Saks spaces (so that the $\pi_{\beta\alpha}$'s are Saks space morphisms), we can define its (Saks space) projective limit $(E, \|\ \|, \tau)$ as follows: as in the first paragraph of this section, we consider the space E of threads as a subspace of $S \prod_{\alpha \in A} E_\alpha$ and give it the induced structure in the sense of 3.5. It can easily be checked that the unit ball of E is τ_∞-closed in $S \prod_{\alpha \in A} E_\alpha$ and so E is complete if each E_α is. We denote this projective limit by $S\text{-}\varprojlim \{E_\alpha, \pi_{\beta\alpha}\}$. Sums and inductive limits of Saks spaces can be defined without difficulty but we shall not require them.

We recall that each Banach space can be regarded as a Saks space - namely the Saks space $(E, \|\ \|, \tau_{\|\ \|})$. The following result shows that the Banach spaces are in a certain sense dense in the Saks spaces and corresponds to the fact that complete locally convex spaces are projective limits of Banach spaces.

3.8. Proposition: A Saks space $(E,\|\ \|,\tau)$ is complete if and only if it is the Saks space projective limit of a system of Banach spaces.

Proof: The sufficiency follows from the remarks above. Necessity: denote by S the family of τ-continuous seminorms p on E which are majorised by $\|\ \|$. Then S is a directed set with the natural (pointwise) ordering and $\|\ \| = \sup S$ (3.1). If $p \in S$ we denote by $\hat{E}_p$ the Banach space associated with p (i.e. the completion of the space E/N_p where $N_p := \{x \in E : p(x) = 0\}$, with the norm induced by p). If $p \le q$, let π_{qp} denote the natural contraction from $\hat{E}_q$ into $\hat{E}_p$. then $\{\pi_{qp} : \hat{E}_q \longrightarrow \hat{E}_p,\ p \le q\}$ is a projective system of Banach spaces and it is not difficult to show that $(E,\|\ \|,\tau)$ is its projective limit.

We remark that if $(E,\|\ \|,\tau)$ is not complete, then the above construction produces its completion in the sense of 3.6. As an example of a Saks space product, consider a family $\{T_\alpha\}_{\alpha\in A}$ of locally compact spaces and let $\mathcal{S}_\alpha$ be a family of subsets of T as in § I.2.D. Denote by T the topological sum of the spaces $\{T_\alpha\}$ and by $\mathcal{S}$ the family $\bigcup_{\alpha\in A} \mathcal{S}_\alpha$ (T_α is regarded as a subspace of T). Then the underlying vector space of $S\prod_{\alpha\in A} C^\infty(T_\alpha)$ can be naturally identified with $C^\infty(T)$ and this induces a Saks space isomorphism between $(C^\infty(T),\|\ \|_\infty,\tau_{\mathcal{S}})$ and $S\prod_{\alpha\in A} C^\infty(T_\alpha)$. This example displays the suitability of a Saks space product in a situation where any locally convex product is hopelessly inadequate.

If T is a locally compact space, then

$$\{\rho_{K_1,K} : C(K_1) \longrightarrow C(K),\ K,K_1 \in \mathcal{K}(T),\ K \subseteq K_1\}$$

(where C(K) denotes the space of continuous, complex-valued functions on K and $\rho_{K_1,K}$ is the restriction operator) is a projective spectrum of Banach spaces and its Saks space projective limit is $(C^\infty(T), \|\ \|_\infty, \tau_K)$.

3.9. Duality: The dual of $(E,\|\ \|,\tau)$ is defined to be the linear span of the set of morphisms from E into $\mathbb{C}$ i.e. it is the dual of the locally convex space $(E,\gamma[\|\ \|,\tau])$ - we denote it by E'_γ. It is a Banach space. Suppose now that E is the Saks space projective limit of the spectrum

$$\{\pi_{\beta\alpha} : E_\beta \longrightarrow E_\alpha,\ \alpha \leq \beta,\ \alpha,\beta \in A\}$$

of Saks spaces. We assume, in addition, that $\pi_\alpha(B_{\|\ \|})$ is τ_α-dense in $B_{\|\ \|_\alpha}$ for each α (π_α is the natural morphism from E into E_α). This condition is satisfied, for example, by the canonical representation of E (3.8). Then each $(E_\alpha)'$ can be regarded as a Banach subspace of $(E,\|\ \|)'$ and the natural injection $i_{\alpha\beta}$ from $(E_\alpha)'_\gamma$ into $(E_\beta)'_\gamma$ $(\alpha \leq \beta)$ is the transpose of $\pi_{\beta\alpha}$.

3.10. Proposition: (a) E'_γ is the Banach space inductive limit of the spectrum

$$\{i_{\alpha\beta} : (E_\alpha)'_\gamma \longrightarrow (E_\beta)'_\gamma,\ \alpha \leq \beta,\ \alpha,\beta \in A\}.$$

(b) a subset H of $(E,\|\ \|)'$ is γ-equicontinuous if and only if there is a sequence (α_n) with values in A and, for each $n \in \mathbb{N}$, a subset H_n of $(E, \| \|)'$ so that

(i) H_n is τ_{α_n}-equicontinuous;

(ii) $\sum_n \sup\{\|f\| : f \in H_n\} < \infty$;

(iii) $H \subseteq \sum H_n$ (i.e. if $f \in H$, $f = \sum_{n=1}^{\infty} f_n$ where $f_n \in H_n$).

Proof: (a) By a standard result on the duals of locally convex projective limits (see SCHAEFER [61], § IV.4.4), the dual of (E,τ) is the subspace $\bigcup_{\alpha \in A} (E_\alpha, \tau_\alpha)'$ of $(E, \| \|)'$. Hence by 1.18(ii) and the remarks at the beginning of this section, E'_γ is the inductive limit of the Banach spaces $\{(E_\alpha)'_\gamma\}$.

(b) We note firstly that if H' is a τ-equicontinuous subset of E' then there is an $\alpha \in A$ so that $H' \subseteq E'_\alpha$ and H' is τ_α-equicontinuous. The sufficiency of the condition follows then from 1.22. On the other hand, if H is γ-equicontinuous then there are α_0, α_1 in A ($\alpha_0 < \alpha_1$) and H_0 τ_{α_0}-equicontinuous in E' (resp. $\tilde{H}_1$ τ_{α_1}-equicontinuous in E') so that

$$H \subseteq H_0 + \varepsilon/2\, B, \quad H \subseteq H_1 + \varepsilon/2\, B$$

($\varepsilon > 0$, B the unit ball of E')

Then if $h \in H$, it has a representations

$$h_0 + \varepsilon/2\, b_0; \quad h_1 + \varepsilon/2\, b_1 \quad (h_0 \in H_0,\ h_1 \in H_1,\ b_0, b_1 \in B)$$

Then $h = h_0 + (h_1 - h_0) + \varepsilon/2\, b_0$ and $\|h_1 - h_0\| \le \varepsilon$.

We define H_1 to be the set of all such $(h_0 - h_1)$ required in the representations of the elements of H. Continuing inductively, we can construct a sequence (H_n) with the required properties.

3.11. Spaces of linear mappings: Let $(E,\| \|,\tau)$ and $(F,\| \|_1,\tau_1)$ be Saks spaces and denote by $L(E,F)$ the space of γ-continuous linear mappings from E into F. Let Σ be a saturated family of norm-bounded sets in E. Then on $L(E,F)$ we consider the following structures:

$\| \|$ - the uniform norm;

τ_Σ - the topology of uniform convergence (for τ_1) on the sets of Σ.

Then $(L(E,F),\| \|,\tau_\Sigma)$ is a Saks space. It is complete if $(F,\| \|_1,\tau_1)$ is and if Σ is the family of all bounded sets in E. It is a Banach space if F is Banach.

3.12. Tensor products: Let $(E,\| \|,\tau)$ and $(F,\| \|_1,\tau_1)$ be Saks spaces. We denote by $E \otimes F$ the algebraic tensor product of E and F. On $E \otimes F$ we consider the following structures:

$$\| \|_\otimes : x \longmapsto \sup \{|(f \otimes g)(x)| : f \in E'_\tau,\ \|f\| \le 1,\ g \in F'_{\tau_1},\ \|g\| \le 1\}$$

(here we are using the norm of f (resp. g) in E'_γ (resp. in F'_γ)).

$\tau \hat{\otimes} \tau_1$: the projective tensor product of τ and τ_1;

$\tau \hat{\hat{\otimes}} \tau_1$: the inductive tensor product of τ and τ_1.

Then $(E \otimes F,\| \|_\otimes,\tau \hat{\otimes} \tau_1)$ and $(E \otimes F,\| \|_\otimes,\tau \hat{\hat{\otimes}} \tau_1)$ are Saks spaces (this follows from 3.1 since $|f \otimes g|$ is continuous for $\tau \hat{\otimes} \tau_1$ and $\tau \hat{\hat{\otimes}} \tau_1$). We denote their completions by $(E \hat{\otimes}_\gamma F,\| \|,\tau \hat{\otimes} \tau_1)$ and $(E \hat{\hat{\otimes}}_\gamma F,\| \|,\tau \hat{\hat{\otimes}} \tau_1)$ resp. Using the representation of the completion by projective limits, we can

give another construction of the tensor product $E \hat{\hat{\otimes}}_\gamma F$:
Let

$$\{\pi_{qp} : \hat{E}_q \longrightarrow \hat{E}_p,\ p,q \in S,\ p \le q\}$$
$$\{\pi_{q_1p_1} : \hat{F}_{q_1} \longrightarrow \hat{F}_{p_1},\ p_1,q_1 \in S_1,\ p_1 \le q_1\}$$

be the canonical representations of $\hat{E}$ and $\hat{F}$. Then

$$\{\pi_{qp} \hat{\otimes} \pi_{q_1p_1} : E_q \hat{\hat{\otimes}} F_{q_1} \longrightarrow E_p \hat{\hat{\otimes}} F_{p_1},\ p \le q,\ p_1 \le q_1\}$$

is a projective spectrum of Banach spaces and its Saks space projective limit is naturally isomorphic to $(E \hat{\hat{\otimes}}_\gamma F, \|\ \|, \tau \hat{\hat{\otimes}} \tau_1)$.

We finish this section with some brief comments on spaces which generalise the class of Banach algebras (resp. C*-algebras) exactly as the Saks spaces generalise the class of Banach spaces.

3.13. <u>Definition</u>: Let A be an algebra with unit e.
A <u>submultiplicative seminorm</u> on A is a seminorm p with the properties $p(xy) \le p(x)p(y)$ $(x,y \in A)$ and $p(e) = 1$;
If A has an involution $x \longrightarrow x^*$, p is a C*-<u>seminorm</u> if, in addition, it satisfies the condition $p(x^*x) = \{p(x)\}^2$ $(x \in A)$.
A <u>Saks algebra</u> is a triple $(A, \|\ \|, \tau)$ where $(A, \|\ \|)$ is a Banach algebra with unit and the locally convex topology can be defined by a family S of submultiplicative seminorms so that $\|\ \| = \sup S$.
A <u>Saks C*-algebra</u> is defined in exactly the same way with the additional requirements that A have an involution and the seminorms of S be C*-seminorms. It follows from this condition that $(A, \|\ \|)$ is then a C*-algebra.

If p is a submultiplicative seminorm, then $\hat{A}_p$ (as defined in the proof of 3.8) has a natural Banach algebra structure. Hence a Saks algebra A has a canonical representation as a projective limit of a spectrum

$$\{\pi_{qp} : \hat{A}_q \longrightarrow \hat{A}_p,\ p \leq q,\ p,q \in S\}$$

where each $\hat{A}_p$ is a Banach algebra and the linking mappings are unit-preserving homomorphisms. Similary, a Saks C*-algebra has a representation with C*-algebras as components and C*-algebra homomorphisms as linking mappings.

3.14. The spectrum: If $(A,\|\ \|,\tau)$ is a Saks algebra, we denote by $M_\gamma(A)$ the set of γ-continuous multiplicative functionals f from A into $\mathbb{C}$ with $f(e) = 1$. $M_\gamma(A)$ is called the spectrum of A. We regard $M_\gamma(A)$ as a topological space with the weak topology induced from $(A', \sigma(A',A))$. Then $M_\gamma(A)$, as a subspace of a locally convex space, is completely regular. $M_\gamma(A)$ is a (topological) subspace of the spectrum of the Banach algebra $(A,\|\ \|)$.

1.4. SPECIAL RESULTS

In 1.5 we saw that the mixed topology could be characterised as the finest linear topology (and hence also as the finest locally convex topology) which agrees with τ on the sets of $\mathcal{B}$.

In general, it is not the finest topology which satisfies this condition. In the Corollary to the following Proposition, we give sufficient conditions for this to be the case.

4.1. Proposition: Let $(E,\mathcal{B},\tau)$ be as in 1.4. Then the following are equivalent:

(a) $\mathcal{B}$ has a basis of τ-compact sets;

(b) there is a Fréchet space F so that $E = F'$, with the structure described in 2.A.

Proof: (b) $\Longrightarrow$ (a) follows from the BOURBAKI-ALAOGLU theorem. (a) $\Longrightarrow$ (b): since (E,γ) is semi-reflexive (1.13), we can identify E with the dual of the Fréchet space E'_γ. Then $\mathcal{B}$ is the equicontinuous bornology of E and $\tau = \sigma(E,E')$ on the sets of $\mathcal{B}$ by compactness.

4.2. Corollary: If (a) or (b) is satisfied then:

(c) $\gamma[\mathcal{B},\tau]$ is the finest topology on E which agrees with τ on the sets of $\mathcal{B}$;

(d) a subset A of E is γ-closed (resp. open) if and only if $A \cap B$ is $\tau|_B$-closed (resp. open) for each $B \in \mathcal{B}$.

Proof: In view of 4.1, this is a restatement of the Banach-Dieudonné theorem ([31], p.274).

We remark that a famous counter-example of GROTHENDIECK ([26], pp.98-99) which we shall not reproduce here shows that 4.2(c) does not hold in general. See also PERROT [47].

4.3. <u>Corollary</u>: Let $(E,\|\ \|,\tau)$ be a Saks space in which $B_{\|\ \|}$ is τ-compact and let A be a subset of E so that $\lambda A = A$ $(\lambda > 0)$ (in particular, if A is a subspace). Then A is γ-closed if and only if $A \cap B_{\|\ \|}$ is τ-closed.

<u>Proof</u>: To check that A is γ-closed, we need only verify that $A \cap nB_{\|\ \|}$ is τ-closed for each $n \in \mathbb{N}$. But $A \cap nB_{\|\ \|} = n(A \cap B_{\|\ \|})$.

Now suppose that we have a space $(E,\mathcal{B},\tau)$ and that F is a vector subspace of E. Then $\mathcal{B}$ and τ induce a bornology $\mathcal{B}_F$ and a topology τ_F on F ($\mathcal{B}_F$ is generated by the balls of $\mathcal{B}$ which are contained in F). An important question is the following: do we have the equality

$$\gamma[\mathcal{B},\tau]|_F = \gamma[\mathcal{B}_F,\tau_F]$$

We clearly have

$$\gamma[\mathcal{B},\tau]|_F \subseteq \gamma[\mathcal{B}_F,\tau_F]$$

(for $(\sum_{k=1}^{n} U_k \cap B_k) \cap F \supseteq \sum_{k=1}^{n} (U_k \cap F) \cap (B_k \cap F))$.

However, there is a counter-example which shows that equality does not hold in general (ALEXIEWICZ and SEMADENI [8], p.133) but the following Proposition gives sufficient conditions for equality:

4.4. Proposition: Suppose that $(E,\mathcal{B},\tau)$ is such that (E,γ) is semi-reflexive and let F be a γ-closed subspace. Then $\gamma[\mathcal{B},\tau]|_F = \gamma[\mathcal{B}_F,\tau_F]$.

Proof: Consider the Fréchet space $G := E'_\gamma$. Then $E = G'$. Since F is $\sigma(E,G)$-closed, F is the dual of the quotient space G/F^o (F^o is the polar of F in G). Hence it suffices to show that every γ-equicontinuous set H in G/F^o is the image, under the quotient mapping π, of a γ-equicontinuous subset of G (note that G/F^o is the dual of F under both of the topologies under consideration). Now we can choose a neighbourhood basis (U_n) of 0 in G so that $\pi(U_n)$ is a neighbourhood basis of zero in G/F^o (since both are Fréchet spaces). Then by 1.22 there is, for each n, a τ-equicontinuous subset H_n of G/F^o so that

$$H \subseteq H_n + \pi(U_n).$$

We can extend H_n to a τ-equicontinuous subset $\widetilde{H}_n$ of G. Then $\widetilde{H} := \cap(\widetilde{H}_n + U_n)$ is a γ-equicontinuous subset of F and $H \subseteq \pi(\widetilde{H})$. For

$$\begin{aligned}\pi^{-1}(H) &\subseteq \pi^{-1}(\cap(H_n + \pi(U_n))) \\ &= \cap\,\pi^{-1}(H_n + \pi(U_n)) \\ &\subseteq \cap(\widetilde{H}_n + U_n + F^o)\end{aligned}$$

and so

$$\begin{aligned}H = \pi(\pi^{-1}(H)) &\subseteq \pi(\cap(\widetilde{H}_n + U_n + F^o)) \\ &\subseteq (\cap(H_n + \pi(U_n))).\end{aligned}$$

We have defined the mixed topology in 1.4 by displaying a neighbourhood basis of zero. However, it is often convenient to characterise a locally convex topology by its continuous seminorms. It is not too difficult to show that the topology γ can be defined by the following seminorms: choose for each $n \in \mathbb{N}$ a τ-continuous seminorm p_n and define, for $x \in E$

$$p(x) := \inf \{ \Sigma\, p_k(x_k)\}$$

where the infimum is taken over all representations of x in the form $x_1 + \ldots + x_n$ where $x_k \in B_k$. Then the set of all such seminorms determines γ (cf. DE WILDE [74], where a corresponding representation for seminorms on inductive limits of locally convex spaces is given). This characterisation is not very useful for applications and we shall now show that in certain cases a much simpler representation for the γ-continuous seminorms on a Saks space can be given. Let $(E,\|\ \|,\tau)$ be a Saks space and let S be a defining family of seminorms for τ which is closed for finite suprema and is such that $\|\ \| = \sup S$. Then for each pair (p_n) and (λ_n) of sequences (where $p_n \in S$ and (λ_n) is a sequence of positive numbers with $\lambda_n \uparrow \infty$),

$$\tilde{p} : x \longrightarrow \sup \lambda_n^{-1}\, p_n(x)$$

is a seminorm on E. The family of all such seminorms defines a locally convex topology $\tilde{\tilde{\gamma}}[\|\ \|,\tau]$ on E.

4.5. <u>Proposition</u>: Let $(E,\|\ \|,\tau)$ be a Saks space and suppose that either

(a) for every $x \in E$, $\epsilon > 0$, $p \in S$, there are elements $x,y \in E$ so that $x = y+z$, $p(z) = 0$ and $\|y\| \le p(x) + \epsilon$;

or (b) $B_{\| \|}$ is τ-compact.

Then $\gamma[\| \|,\tau] = \tilde{\tilde{\gamma}}[\| \|,\tau]$.

Proof: We verify firstly the inclusion

$$\gamma[\| \|,\tau] \supseteq \tilde{\tilde{\gamma}}[\| \|,\tau]$$

which is valid without assumptions (a) or (b). We show that $\tilde{\tilde{\gamma}}[\| \|,\tau]$ is coarser than τ on $B_{\| \|}$ (this suffices by 1.7). A $\tilde{\tilde{\gamma}}$-neighbourhood of zero in $B_{\| \|}$ has the form

$$B_{\| \|} \cap \bigcap_{n=1}^{\infty} \{x : p_n(x) \le \lambda_n\}$$

where (p_n) and (λ_n) are as above. But if $N \in \mathbb{N}$ is chosen so that $\lambda_n > 1$ for $n \ge N$, then

$$B_{\| \|} \cap \bigcap_{n=1}^{\infty} \{x : p_n(x) \le \lambda_n\} = B_{\| \|} \cap \bigcap_{n=1}^{N} \{x : p_n(x) \le \lambda_n\}$$

and the latter is a τ-neighbourhood of zero in $B_{\| \|}$.

We now consider the reverse inclusion under assumption (a). Every γ-neighbourhood of zero contains a set of the form

$$U_0 \cap \bigcap_{n=1}^{\infty} (U_n + nB_{\| \|})$$

where $U_n := \{x : p_n(x) \le \varepsilon_n\}$ $(\varepsilon_n > 0,\ p_n \in S)$ (1.16).
We can assume, in addition, that $p_n \le p_{n+1}$ for each n.
Put $\lambda_0 := \min(\varepsilon_0, 1/2)$, $\lambda_n := n/2$ $(n \ge 1)$. Choose $x \in E$ with $\tilde{p}(x) \le 1$ where $\tilde{p} := \sup_n \lambda_n^{-1} p_n$. Then for any n there is a decomposition $x = y+z$ of x where $\|y\| \le p_n(x) + (n/2)$ and $p_n(z) = 0$. Then $z \in U_n$ and $y \in nB_{\| \|}$ so that $x \in U + nB$
Hence $U_0 \cap \bigcap_{n=1}^{\infty} (U_n + nB_n) \supseteq \{x : \tilde{p}(x) \le 1\}$.

Now we assume that $B_{\| \|}$ is τ-compact. Let U be a γ-open neighbourhood of zero. Then there is a $p_o \in S$ and an $\varepsilon > 0$ so that $U \cap B_{\| \|} \supseteq \{x : p_o(x) \leq \varepsilon\} \cap B_{\| \|}$.

Suppose that we can find $p_1, \ldots, p_n \in S$ so that

$$\bigcap_{k=1}^{n} \{x : p_k(x) \leq \lambda_k\} \cap nB_{\| \|} \subseteq U \cap nB$$

where $\lambda_o = \varepsilon$, $\lambda_n = (n-1)$ $(n > 1)$. We shall prove the existence of a $p_{n+1} \in S$ so that

$$\bigcap_{k=1}^{n+1} \{x : p_k(x) \leq \lambda_k\} \cap (n+1)B_{\| \|} \subseteq U \cap (n+1)B_{\| \|}.$$

Then we can construct, by induction, a sequence (p_n) in S so that

$$\bigcap_{k=1}^{\infty} \{x : p_k(x) \leq \lambda_k\} \cap nB_{\| \|} \subseteq U$$

for each n and so

$$\bigcap_{k=1}^{\infty} \{x : p_k(x) \leq \lambda_k\} \subseteq U$$

which proves the result.

To prove the existence of p_{n+1} we argue by contradiction. If no such seminorm exists, then

$$C_q := \bigcap_{k=1}^{n} \{x : p_k(x) \leq \lambda_k\} \cap \{x : q(x) \leq n\}$$

has non-empty intersection with the τ-compact set $(n+1)B_{\| \|} \setminus U$ for each $q \in S$. Hence, by the finite intersection property, there is an $x_o \in (n+1)B_{\| \|} \setminus U \cap \bigcap_{q\in S} C_q$. Hence $q(x_o) \leq n$ for each $q \in S$ and so $\|x_o\| \leq n$. Then

$$x_o \in U \cap nB_{\| \|} \subseteq U \cap (n+1)B_{\| \|}$$

which is a contradiction.

Using this result, we can obtain new sufficient conditions for the equality

$$\gamma[\|\ \|_F,\tau_F] = \gamma[\|\ \|,\tau]|_F$$

to hold for a subspace F of a Saks space $(E,\|\ \|,\tau)$ (cf. 4.4).

4.6. <u>Proposition</u>: Let F be a subspace of a Saks space $(E,\|\ \|,\tau)$ and suppose that $(F,\|\ \|_F,\tau_F)$ satisfies (a) or (b) of 4.5. Then

$$\gamma[\|\ \|_F,\tau_F] = \gamma[\|\ \|,\tau]|_F$$

<u>Proof</u>: It is sufficient to show that a $\gamma[\|\ \|_F,\tau_F]$-continuous seminorm of the form $p := \sup_{n\in N} \lambda_n^{-1} p_n$ on F is the restriction of a $\gamma[\|\ \|,\tau]$-continuous seminorm $\tilde{p}$ on E. But if $\tilde{p}_n$ is an extension of p_n to a τ-continuous seminorm on E then $\tilde{p} := \sup_{n\in N} \lambda_n^{-1} \tilde{p}_n$ has the required property.

4.7. <u>Corollary</u>: Let $(E,\|\ \|,\tau)$ be a Saks space, F a subspace so that the unit ball of $(F,\|\ \|_F)$ is τ_F-compact. Then the restriction mapping induces an isometry from $(E,\gamma)'/F^0$ onto $(F,\gamma)'$.

<u>Proof</u>: The restriction mapping from $(E,\gamma)'$ into $(F,\gamma)'$ is obviously a norm-contraction and it is surjective by 4.6. We must show that if $f \in (F,\gamma)'$ with $\|f\| = 1$ and if $\varepsilon > 0$, then there is an extension $\tilde{f}$ of f to an element of $(E,\gamma)'$ with $\|\tilde{f}\| \leq 1+\varepsilon$. Since $B_{\|\ \|_F}$ is τ_F-compact, there is an

absolutely convex γ-neighbourhood U of zero so that

$$W := (\frac{1}{1+\varepsilon} B_{\| \|_F}) + U \subseteq \{x \in F : |f(x)| \leq 1\}$$

We can extend the Minkowski functional of W to a γ-continuous seminorm $\widetilde{p}$ on E so that

$$\frac{1}{1+\varepsilon} \widetilde{B}_{\| \|} \subseteq \{x : \widetilde{p}(x) \leq 1\}.$$

The result then follows by applying the Hahn-Banach theorem to f to obtain a linear form $\widetilde{f}$ on E which extends f and is majorised by $\widetilde{p}$.

Now we discuss a theorem of Banach-Steinhaus type for mixed topologies. Since such spaces are not, in general, barrelled the classical approach cannot be employed. Indeed, if one considers the identity mapping from $(\ell^1, \| \|_1, \tau_s)$ into $(\ell^1, \| \|)$ which is not γ-continuous, but is the pointwise limit of the projection mappings

$$p_n : (x_k) \longrightarrow (x_1, \ldots, x_n, 0, 0, \ldots)$$

which are continuous, it is clear that a Banach-Steinhaus theorem can only hold under rather special restrictions.

4.8. Definition: Let E be a locally convex space. Then E has the Banach-Steinhaus property if for every sequence (T_n) of continuous linear mappings from E into a locally convex space F such that the pointwise limit

$$T : x \longmapsto \lim (T_n x) \quad (x \in E)$$

exists, this limit is continuous. The following remarks are evident:

a) in this definition, one can assume that F is a Banach space;

b) inductive limits and quotients of spaces with the Banach-Steinhaus property also have this property;

c) the Banach-Steinhaus property is possessed by σ-barrelled spaces (and in particular by barrelled spaces) (recall that a locally convex space is σ-<u>barrelled</u> if every barrel which is the intersection of a countable family of neighbourhoods of zero is itself a neighbourhood of zero);

d) if E is a Mackey space, it is sufficient to verify the condition for linear forms (i.e.that E' is $\sigma(E',E)$ sequentially complete).

4.9. <u>Definition</u>: Let $(E,\mathcal{B},\tau)$ be as in 1.4. We say that E satisfies <u>condition Σ_1</u> if, for every $n \in \mathbb{N}$ every $x_0 \in B_n$, and every τ-neighbourhood of zero U there is a τ-neighbourhood V of zero so that

$$V \cap B_n \subseteq ((x_0 + U) \cap B_n) - ((x_0 + U) \cap B_n).$$

The importance of this property is that it allows us to deduce $\tau|_B$-continuity of an operator on some $B \in \mathcal{B}$ from its $\tau|_B$-continuity at a single point.

4.10. <u>Lemma</u>: Let $(E,\mathcal{B},\tau)$ satisfy condition Σ_1 and let $\mathcal{F}$ be a family of linear mappings from E into a locally convex space F. Suppose that the following condition is satisfied:

for each $B \in \mathcal{B}$ and each absolutely convex neighbourhood W of 0 in F there is an $x_0 \in B$ and a τ-neighbourhood U of 0 so that $T((x_0 + U) \cap B) \subseteq Tx_0 + W$. Then $\mathcal{F}$ is γ-equicontinuous.

Proof: By Grothendieck's Lemma (1.8) we need only show that $\mathcal{F}|_B$ is $\tau|_B$-equicontinuous at 0. Let W be an absolutely convex neighbourhood of zero in F. Then there is a τ-neighbourhood U of zero so that

$$T((x_0 + U) \cap B) \subseteq Tx_0 + W/2$$

for each $T \in \mathcal{F}$. Choose V as in 4.9. Then for each $T \in \mathcal{F}$,

$$\begin{aligned} T(V \cap B) &\subseteq T((x_0 + U) \cap B - (x_0 + U) \cap B) \\ &\subseteq (Tx_0 + W/2) - (Tx_0 + W/2) \\ &= W. \end{aligned}$$

4.11. Proposition: Suppose

a) that $(E, \mathcal{B}, \tau)$ satisfies condition Σ_1;

b) $\mathcal{B}$ has a basis $\mathcal{B}_1$ so that $(B, \tau|_B)$ is a Baire space for each $B \in \mathcal{B}_1$.

Then (E, γ) has the Banach-Steinhaus property.

Proof: Let (T_n) be a pointwise convergent sequence of γ-continuous linear mappings from E into a Banach space F. Let $B \in \mathcal{B}$ be a $\tau|_B$-Baire space. We shall find a point $x_0 \in B$ and a τ-neighbourhood U of zero so that

$$T_n((x_0 + U) \cap B) \subseteq T_n(x_0) + \varepsilon B_{\| \|_F} \quad \text{for each } n.$$

Then $\{T_n\}$ will be γ-equicontinuous by 4.10 and this will suffice to prove the result. Let

$$A_n := \{x \in B : \|T_p x - T_q x\| \le \varepsilon/3 \quad \text{for } p,q \ge n\}$$

Then A_n is $\tau|_B$-closed and $B = \bigcup_{n \in N} A_n$. Hence there is an $n_o \in \mathbb{N}$ so that A_{n_o} contains an interior point x_o. Now choose a τ-neighbourhood U of O so that $(x_o + U) \cap B \subseteq A_{n_o}$ and $\|T_{n_o}x - T_{n_o}x_o\| \le \varepsilon/3$ if $x \in (x_o + U) \cap B$. Then if $x \in (x_o + U) \cap B$

$$\|T_n x - T_{n_o}x\| \le \|T_n x - T_{n_o}x\| + \|T_{n_o}x - T_{n_o}x_o\| + \|T_{n_o}x_o - T_n x_o\| \le \varepsilon$$

4.12. Corollary: Suppose a) that $(E,\mathcal{B},\tau)$ satisfies Σ_1 and b) $\mathcal{B}$ has a basis of $\tau|_B$-compact or $\tau|_B$-metrisable and complete sets. Then (E,γ) has the Banach-Steinhaus property.

One problem which has proved to be central in the theory of function spaces with strict topologies is that of determining whether the given space is a Mackey space (i.e. if it has the finest locally convex topology compatible with its dual). Most familiar Mackey spaces have this property by virtue of some stronger property (e.g. that of being bornological). Once again, we must seek another approach for mixed topologies. Our first result gives a positive answer for Saks space products of Banach spaces. We require the following preparatory result.

4.13. <u>Proposition</u>: Let $\{(E_\alpha, \| \|_\alpha)\}_{\alpha \varepsilon A}$ be a family of Banach spaces. Then

1) $(S \Pi E_\alpha, B \Sigma E'_\alpha)$ is a dual pair under the bilinear mapping

$$((x_\alpha),(f_\alpha)) \longmapsto \sum_{\alpha \varepsilon A} f_\alpha(x_\alpha).$$

2) Under this duality, $B \Sigma E'_\alpha$, with its Banach space structure is identifiable with the γ-dual of $S \Pi E_\alpha$;

3) a subset C of $B \Sigma E'_\alpha$ is γ-equicontinuous if and only if it is norm-bounded and, for each $\varepsilon > 0$, there is a finite subset J of A so that $\sum_{\alpha \varepsilon A \setminus J} \|f_\alpha\| \leq \varepsilon$ for each $(f_\alpha) \varepsilon C$.

<u>Proof</u>: We can regard $S \Pi E_\alpha$ as the projective limit of the spectrum defined by the spaces $\{B \prod_{\alpha \varepsilon J} E_\alpha : J \varepsilon \mathcal{J}(A)\}$ where $\mathcal{J}(A)$ denoted the family of finite subsets of A. Now there is a natural isometry from $(B \prod_{\alpha \varepsilon J} E_\alpha)'$ onto $B \sum_{\alpha \varepsilon J} E'_\alpha$ and the result follows then from 3.10.

We shall require the following version of Schur's theorem:

4.14. <u>Proposition</u>: A subset C of $\ell^1(S)$ is relatively $\sigma(\ell^1, \ell^\infty)$-compact if and only if it is norm-bounded and for each $\varepsilon > 0$ there is a finite subset J of S so that $\sum_{\alpha \varepsilon S \setminus J} \|x_\alpha\| \leq \varepsilon$ for each $(x_\alpha) \varepsilon C$.
In particular, there is a countable subset S_1 of S so that $C \subseteq \ell^1(S_1)$.

4.15. Proposition: Let $\{E_\alpha\}$ be as in 4.13 with $E := S\prod E_\alpha$. Then (E,γ) is a Mackey space (i.e. $\gamma = \tau(E,E'_\gamma)$).

Proof: It is sufficient to show that if C is a weakly compact subset of $B\sum E'_\alpha$, then C satisfies condition 4.13.3). We first show that the support A_1 of C is countable. (A_1 is the set of those $\beta \in A$ for which an $(f_\alpha) \in C$ exists with $f_\beta \neq 0$). For each $\beta \in A_1$ we choose an $x_\beta \in E_\beta$ so that $\|x_\beta\|_\beta = 1$ and $f_\beta(x_\beta) \neq 0$ for some $(f_\alpha) \in C$. Then the mapping

$$(f_\alpha) \longmapsto (f_\alpha(x_\alpha))$$

from $B\sum E'_\alpha$ into $\ell^1(A)$ is $\sigma(B\sum E'_\alpha, B\prod E_\alpha) - \sigma(\ell^1(A),\ell^\infty(A))$ continuous. For this is equivalent to the fact that for each $(\lambda_\alpha) \in \ell^\infty(A)$ the linear form

$$(f_\alpha) \longrightarrow \sum f_\alpha(x_\alpha)\lambda_\alpha$$

on $B\sum E'_\alpha$ is defined by an element of $B\prod E_\alpha$ - which it clearly is - the element $(\lambda_\alpha x_\alpha)$. The image of C under this mapping is thus weakly compact and so has countable support (4.14). But, by construction, the support of this set is A_1. Thus we can consider the case where the indexing set is $\mathbb{N}$. If C does not satisfy the condition of 4.13.3), there is a strictly increasing sequence of positive integers (n_k) and a sequence $(f^{(k)})$ in C so that $\sum_{n=n_k+1}^{n_{k+1}} \|f_n{}^k\| \geq \epsilon$ for some positive ϵ. For each n, there is an $x_n \in E_n$ so that $\|x_n\| = 1$ and $f_n^{(k)}(x_n) \geq \|f_n^{(k)}\|/2$ $(n_k < n \leq n_{k+1})$.

Then if $x := (x_n)$, we have

$$\sum_{n>n_k} |f_n^{(k)}(x_n)| \geq \sum_{n=n_k+1}^{n_{k+1}} |f_n^{(k)}(x_n)| \geq \varepsilon/2$$

and so $\{(f_n(x_n)) : f \in C\}$ is not weakly compact (4.14) - contradiction.

4.16. Proposition: Let $\{(E_\alpha, \|\ \|_\alpha)\}_{\alpha\in A}$ be a family of Banach spaces and let $(E, \|\ \|, \tau)$ be the product $S \prod_{\alpha\in A} E_\alpha$. Then (E,γ) has the Banach-Steinhaus property.

Proof: By 4.15 and 4.8.d), it is sufficient to show that the pointwise limit of a sequence (f_n) of continuous linear forms is continuous. We use the isometry $E'_\gamma = B \sum_{\alpha\in A} E'_\alpha$. Let $f_n = (f_n^\alpha)$. Then the sequence $(f_n^\alpha)_{\alpha\in A}$ is pointwise convergent to a continuous linear form f^α on E_α for each $\alpha \in A$. By the principle of uniform boundedness, $\sup\{\|f_n\| : n \in \mathbb{N}\} < \infty$ and so $\sup\{\sum_{\alpha\in A} \|f_n\| : n \in \mathbb{N}\} < \infty$.

Hence $\sum_{\alpha\in A} \|f^\alpha\| < \infty$ and so there is a γ-continuous linear form f on E with $f|_{E_\alpha} = f^\alpha$ for each $\alpha \in A$. A simple $\varepsilon/3$ argument shows that f is the pointwise limit of the sequence (f_n).

In 4.15 and 4.16 we have shown that Saks spaces of a very special type have two important properties. Every Saks space is a closed subspace of such a space. However, these properties are not inherited by closed subspaces. We shall now show that under certain circumstances a Saks space is a complemented subspace of a Saks space product of Banach spaces - in this case the properties are inherited.

4.17. Definition: Let $\{\pi_{\beta\alpha} : E_\beta \longrightarrow E_\alpha, \alpha,\beta \in A, \alpha \leq \beta\}$ be a projective system of Banach spaces, $(E,\|\ \|,\tau)$ its Saks space projective limit and π_α the natural mappings from E into E_α. A partition of unity is a family $\{T_\alpha\}$ of norm contractions where $T_\alpha : E_\alpha \longrightarrow E$ so that

a) for each $\beta \in A$, $\{\alpha \in A : \pi_\beta \circ T_\alpha \neq 0\}$ is finite;

b) for each $x \in S\Pi E_\alpha$, $\beta \in A$ $\|\sum_{\alpha\in J} \pi_\beta T_\alpha(x_\alpha)\| \leq \|x\|$;

c) for each $x \in E$, $\sum_{\alpha\in A} T_\alpha \circ \pi_\alpha(x_\alpha) = x$ (convergence in τ).

4.18. Proposition: If a partition of unity exists then (E,γ) is a complemented subspace of $(S\prod_{\alpha\in A} E_\alpha,\gamma)$.

Proof: We shall show that the natural injection $x \longmapsto (\pi_\alpha(x))_{\alpha\in A}$ which is γ-continuous from E into $S\Pi E_\alpha$ has a left inverse. In fact the mapping $T : (x_\alpha) \longmapsto \Sigma\, T_\alpha x_\alpha$ is a left inverse. Firstly, this mapping is well-defined. For the right hand side converges (we need only show that $\sum_{\alpha\in A} \pi_\beta \circ T_\alpha(x_\alpha)$ converges in E_β for each β and this follows from a)). Secondly, it is a norm contraction by b) and is τ-continuous on the unit ball of $S\prod_{\alpha\in A} E$ - once more by b). It is a right inverse by c).

4.19. Corollary: If a partition of unity exists then

1) (E,γ) is a Mackey space;

2) (E,γ) has the Banach-Steinhaus property.

We recall that a locally convex space has the approximation property if the identity mapping is uniformly approximable on compact sets by continuous linear mappings of finite rank. A normed space has the metric approximation property if one can, in addition, demand that the approximating operators be contractions. A locally convex space which is the projective limit of a spectrum of normed spaces with the approximation property also has the approximation property. In the following, we give a corresponding result for Saks space projective limits. Unfortunately we require rather strong additional hypotheses but, as we shall see, these are often fulfilled in applications.

4.20. Proposition: Let $\{\pi_{\beta\alpha} : E_\beta \longrightarrow E_\alpha, \alpha,\beta \in A, \alpha \leq \beta\}$ be a projective system of Banach spaces, $(E,\|\ \|,\tau)$ its Saks space projective limit. Suppose that

a) each E_α has the metric approximation property;

b) there is a norm contraction $i_\alpha : E_\alpha \longrightarrow E$ so that $\pi_\alpha \circ i_\alpha = id_{E_\alpha}$ for each $\alpha \in A$.

Then (E,γ) has the approximation property.

Proof: Let K be an absolutely convex compact subset of E and V a τ-neighbourhood of zero (we can and do assume that $2K \subseteq B_{\|\ \|}$).

There is an $\alpha \in A$ so that $V \cap B_{\|\ \|} \supseteq \varepsilon\pi_\alpha^{-1}(B_{\|\ \|_\alpha})$ $(\varepsilon > 0)$. Then there is an operator $T : E_\alpha \longrightarrow E_\alpha$ of finite rank with $\|T\| \leq 1$ and $(id_{E_\alpha} - T)(\pi_\alpha(K)) \subseteq \varepsilon B_{\|\ \|_\alpha}$. Let $\widetilde{T} := i_\alpha \circ T \circ \pi_\alpha$. Then $\widetilde{T}$ is of finite rank and $\|\widetilde{T}\| \leq 1$.

If $x \in K$ then $(id_{E_\alpha} - T)(\pi_\alpha(x)) \in \varepsilon B_{\|\ \|_\alpha}$. Hence

$$\pi_\alpha(id_E(x) - \widetilde{T}x) \in \varepsilon B_{\|\ \|_\alpha} \quad \text{and so} \quad (id_E - \widetilde{T})(K) \subseteq \varepsilon\pi^{-1}(B_{\|\ \|_\alpha}).$$

Since $(id_E - \widetilde{T})(K) \subseteq B_{\|\ \|}$ we have $(id_E - \widetilde{T})(K) \subseteq V$.

4.21. Corollary: Let $\{(E_\alpha, \|\ \|_\alpha)\}_{\alpha \in A}$ be a family of Banach spaces, $(E, \|\ \|, \tau)$ its Saks space product. Then if each E_α has the metric approximation property, (E,γ) has the approximation property.

Proof: We can regard E as the projective limit of the spectrum $\{B \prod_{\alpha \in J} E_\alpha : J \in F(A)\}$ where $F(A)$ denotes the family of finite subsets of A. Now $B \prod_{\alpha \in J} E_\alpha$ has the metric approximation property and 4.20.b) is satisfied.

4.22. Corollary: Let $\{\pi_{\beta\alpha} : E_\beta \longrightarrow E_\alpha, \alpha, \beta \in A, \alpha \leq \beta\}$ be a projective system of Banach spaces, $(E, \|\ \|, \tau)$ its Saks space projective limit. Then if there exists a partition of unity, (E,γ) has the approximation property.

Proof: This follows from 4.18, 4.21 and the fact that a complemented subspace of a locally convex space with the approximation property has itself the approximation property.

Now we turn our attention to closed graph theorems for spaces with mixed topologies. Since the graph of the identity mapping: $(E,\gamma) \longrightarrow (E,\|\ \|)$ is closed for any Saks space it is once again clear that any such closed graph theorem must employ rather special restrictions. Nevertheless, we can obtain two useful closed graph theorems, one based on a closed graph theorem for locally convex spaces due to KALTON and one obtained by combining a closed graph theorem for bornological spaces (due to BUCHWALTER) and a closed graph theorem for topological spaces.

If T is a linear mapping from E into F (E,F locally convex spaces), we put

$$D(T') = \{f \in F' : (x \longrightarrow f(Tx)) \in E'\}$$

i.e. $D(T') = \{f \in F' : T^*f \in E'\}$ (T^* is the algebraic adjoint of T). We note that T is weakly continuous if and only if $D(T') = F'$. A simple calculation shows that

$$(\Gamma(T))^{\circ} = \{(-T'f,f) : f \in D(T')\}$$

where $(\Gamma(T))^{\circ}$ is the polar of the graph $\Gamma(T)$ of T in $E' \times F'$. Now if $x \in (D(T'))^{\circ}$ (polar in F) then $(0,x)$ vanishes on the right hand side of the above equation and so $(0,x) \in (\Gamma(T))^{\circ\circ}$. Hence if T has a closed graph (so that $(\Gamma(T))^{\circ\circ} = \Gamma(T)$) we have $x = T0 = 0$ and so $D(T')^{\circ} = \{0\}$ that is, $D(T')$ is dense in F'.

4.23. Proposition: Let E be a Mackey space for which E' is $\sigma(E',E)$-sequentially complete. Then a linear mapping T from E into a separable Fréchet space F is continuous if and only if it has a closed graph.

Proof: Since E is a Mackey space, it suffices to show that a linear mapping T from E into F is weakly continuous if it has a closed graph. By the above remarks we need only show that D(T') is $\sigma(F',F)$-closed. Since E' is weakly sequentially complete, D(T') is sequentially closed. For if (f_n) is a sequence in D(T') which tends weakly to a linear form f then $(T'(f_n))$ is $\sigma(E',E)$-Cauchy and so converges to a continuous linear form on E. But the limit is of course T^*f and so $f \in D(T')$. By the Banach-Dieudonné theorem, to show that D(T') is closed it suffices to show that $D(T') \cap U^o$ is weakly closed for each neighbourhood U of zero in F. But U^o is weakly metrisable (since F is separable) and so $D(T') \cap U^o$, being sequentially closed in U^o, is closed.

4.24. Proposition: Let $(E,\| \|,\tau)$ be a Saks space and suppose that

a) $(E,\| \|,\tau)$ satisfies the conditions of 4.12;

or b) E is the Saks space projective limit of a spectrum of Banach spaces with partition of unity;

Then a linear mapping T from E into a separable Fréchet space is γ-continuous if and only if its graph is closed.

Proof: Under these conditions, E' is weakly sequentially complete (4.12 and 4.19) and so, by 4.24, T is weakly continuous. We show that it is continuous. By projecting down into the Banach spaces associated to the continuous seminorms of F we can assume, without loss of generality, that F is a separable Banach space. In addition we can assume that F has a Schauder basis (x_k) (since every separable Banach space is a subspace of a Banach space with Schauder basis - e.g. the space $C[0,1]$ - see BANACH [13], pp. 185 and p. 112). Let (P_n) be the associated sequence of projection operators (i.e. $P_n : \sum_{k=1}^{\infty} \xi_k x_k \longrightarrow \sum_{k=1}^{n} \xi_k x_k$). Then $P_n \circ T \longrightarrow T$ pointwise on E and so by 4.12 and 4.19 it will suffice to show that $P_n \circ T$ is γ-continuous. But this is clear since $P_n \circ T$ is weakly continuous and takes values in a finite dimensional subspace of F.

4.25. Corollary: Let $(E,\| \|,\tau)$ be a Saks space as in 4.12 and suppose that (E,γ) is separable. Then (E,γ) is a Mackey space.

Proof: We must show that any weakly continuous mapping from E into a Banach space is continuous. But the range of such a mapping is weakly separable and so separable. Its continuity follows then from 4.24.

4.26. Corollary: Let $(E,\| \|,\tau)$ be as in 4.24 and let (F,β) be a separable locally convex space. Then if $T : E \longrightarrow F$ is a linear mapping which is γ-α continuous where α is a locally convex topology on F so that β has a basis of α-closed sets, T is γ-α continuous.

Proof: The hypothesis on α and β implies that β is the topology of uniform convergence on a family S of subsets of the α-dual of F (bipolar theorem!). Then if $M := \bigcup S$, T is γ-$\sigma(F,M)$-continuous. The continuity of T follows from 4.24 since its graph is γ-β closed (at this stage, we can assume that F is a Banach space).

4.27. Remark: Note that in the proof of 4.24 we have actually proved that a locally convex space with the Banch-Steinhaus property satisfies a closed graph theorem where the range space is a separable Fréchet space. On the other hand, a space with the latter property satisfies a Banach-Steinhaus theorem for mappings into a separable Fréchet space. One may compare this with the fact that a locally convex space is barrelled if and only if it satisfies a closed graph theorem with Banach spaces as range spaces. ([61], IV.8.6).

4.28. Definition: Let $(E,\mathcal{B})$ and $(F,\mathcal{B}_1)$ be bornological spaces. A linear mapping T from E into F has a Mackey closed graph if for each $B \in \mathcal{B}$, $C \in \mathcal{B}_1$,

$$\Gamma(T) \cap (E_B \times F_C)$$

is closed in the normed space $E_B \times F_C$.

4.29. Lemma: Let $(E, \| \ \|)$ be a Banach space, $(F, \mathcal{B})$ a complete bornological space of countable tape, T a linear mapping from E into F with a Mackey closed graph. Then T is bounded.

Proof: Let $\{B_n\}$ be a basis for $\mathcal{B}$ and for each n put

$$E_n := \Gamma(T) \cap (E \times F_{B_n})$$

so that E_n has a natural Banach space structure. Denote by p_n the projection from $E \times F_{B_n}$ into E. Then

$$p_n(E_n) = \{x \in E : Tx \in F_{B_n}\}.$$

Now $E = \bigcup p_n(E_n)$ and hence at least one $p_{n_o}(E_{n_o})$ is non-meagre. Hence by a theorem of BANACH ([13], p. 38), p_{n_o} maps E_{n_o} onto E that is $T(E) \subseteq F_{B_{n_o}}$. Then T is bounded from E into $F_{B_{n_o}}$ by the classical closed graph theorem.

4.30. Proposition: Let $(E, \mathcal{B})$ and $(F, \mathcal{B}_1)$ be complete bornological spaces, F of countable type. Then a linear mapping T from E into F is bounded if and only if it has a Mackey closed graph.

Proof: The boundedness of T is obtained by applying the Lemma to the restriction of T to the Banach spaces $\{E_B : B \in \mathcal{B}\}$.

4.31. Lemma: Let X, Y be Hausdorff topological spaces with Y compact. Then a mapping $f : X \longrightarrow Y$ is continuous if and only if its graph is closed.

Proof: see ČECH [16], p. 799.

4.32. Proposition: Let $(E,\mathcal{B},\tau)$, $(F,\mathcal{B}_1,\tau_1)$ be as in 1.4 and suppose that

a) $(E,\mathcal{B})$ is complete;

b) $\mathcal{B}_1$ has a basis of τ_1-compact sets.

Then a linear mapping from E into F is γ-continuous if and only if it has a closed graph.

Proof: If the graph of T is γ-closed then it is Mackey closed for the bornologies $(\mathcal{B},\mathcal{B}_1)$. Then it is $\mathcal{B}$-$\mathcal{B}_1$-bounded by 4.30. The result now follows by applying 4.31 to the restriction of T to the sets of $\mathcal{B}$.

4.33. Remarks: I. 4.24 states that, under the given hypotheses, (E,γ) belongs to the class of locally convex spaces which satisfy a closed graph theorem with a separable Banach space as range space. KALTON [29] has given the following internal characterisation of this class of spaces (which he denotes by $C(\zeta_B)$): a locally convex space E belongs to $C(\zeta_B)$ if and only if each $\sigma(E',E)$ bounded, metrisable ball in E' is equicontinuous.

Using results of VALDIVIA, MARQUINA [36] has shown further that these spaces satisfy a closed graph theorem where the range space is an ω-WCG Banach space (that is, a Banach space which is the union of countably many weakly compact sets or has a dense subset of this form), in particular, where the range space is reflexive. In addition the words "Banach space" in the above formulation can be replaced by "B_r-complete space" (SCHAEFER [61], p. 162).

II. We remark that if E is an infinite dimensional Saks space, then (E,γ) is never nuclear. For we can assume that (E,γ) is complete (if not, consider the completion which is nuclear). Then (E,γ) is semi-Montel (SCHAEFER [61], p. 101, Cor. 2) and so E is the dual of a Banach space F and γ is the topology $\tau_c(E,F)$ (4.1). Now the latter topology is never nuclear if E is infinite-dimensional (private communication of HOGBE-NLEND - this fact follows from Lemma 1 in SHAPIRO [68]).

The class of nuclear spaces of the form $(E,\gamma[\mathcal{B},\tau])$ coincides with the class of those spaces which are the strong duals of Fréchet nuclear spaces (i.e. the (DFN)-spaces). Hence they are uninteresting from the point of view of mixed topologies.

III. If $(E,\| \|,\tau)$ is a Saks space and F is a subspace of E, we say (following SHAPIRO) that

F is <u>weakly normal</u> if $B_{\| \|_F}$ is $\sigma(E,E'_\gamma)$-compact;

F is <u>normal</u> if $B_{\| \|_F}$ is γ-compact.

Then it follows easily from the results of this section that

1) every normal subspace is weakly normal and every weakly normal subspace is norm-closed;

2) F is weakly normal if and only if it is isometric to the dual of E'_γ/F^0 (under the canonical mapping);

3) F is normal if and only if it is weakly normal and $\gamma|_F$ coincides with $\tau_c(F,E'/F^o)$.

These results are obtained, for subspaces of $(C^\infty(S),\| \|,\tau_K)$ (K the family of compact subsets of S - cf. I.2.D) by SHAPIRO [69].

IV. One of the features of many of the concrete Saks spaces which we shall examine in the following chapters is the fact that they possess bases (in contrast to the associated Banach spaces which are usually not separable). The classical theory of bases for locally convex spaces employs some condition like barrelledness and so is not applicable to Saks spaces. It is curious to note that DE GRANDE-DE KIMPE [25], in extending these results to non-barrelled spaces, introduced the classes of G-<u>spaces</u> i.e. locally convex spaces (E,τ) which satisfy the following conditions:

a) E has the Mackey topology;

b) E' is weakly sequentially complete.

i.e. precisely the class of locally convex spaces which satisfy a closed graph theorem with separable Fréchet spaces as range. As we shall see, the combination of the above properties will recur with unnatural frequency in the study of concrete Saks spaces. In fact, the determination of whether a given Saks space possesses these properties is often one of the most challenging and fruitful aspects of its theory. In view of these remarks we quote without proof some of DE GRANDE-DE KIMPE's results on G-spaces with bases:

A. If (E,τ) is a G-space with a Schauder basis, then τ is the finest locally convex topology on E for which the sequence is a basis;

B. If E is a G-space with a Schauder basis then this basis is equicontinuous (i.e. the associated projection operators are equicontinuous).

C. If (E,τ) is a G-space with a weak Schauder basis, then this is a strong basis (i.e. a basis for the original topology).

The first statement, applied to the space $L^{\infty}([0,1])$ with the mixed topology β_s introduced in III.1, gives the rather curious result that β_s is the finest locally convex topology τ on L^{∞} with the property that the HAAR systems form a basis with respect to τ.

We remark that B and C follow easily from the closed graph theorem (4.23).

I.5. NOTES

I.1 consists essentially of an exposition of the results of WIWEGER [78] - [80]. For the theory of convex bornologies see BUCHWALTER [15] and HOGBE-NLEND [27]. 1.5 identifies the mixed topology as a generalised inductive limit in the sense of GARLING [23]. Most of the results of I.1 are contained in his results. For articles which deal with related topics see NOUREDDINE [37] - [41], PERROT [47],[48],[49], PERSSON [50], PRECUPANU [53], ROELCKE [57],[58], RUESS [59]. COOK and DAZORD [19] have studied mixed structures in the context of limit spaces. STROYAN [70] uses non-standard analysis. The localisation principle involved appears in earlier papers (for examples PTAK's fundamental paper [54] on the closed graph theorem - see also COLLINS [18] and WHEELER [73]).

The completeness theorem of RAIKOV referred to in the paragraph before 1.13 can be found in [55]. The proof used in 1.13 is taken from DE WILDE and HOUET [76],[77].

<u>I.2.</u> Examples A,B,C,D,E are classical and can be found at various levels of generality in the first papers on Saks spaces, two-norm spaces and mixed topologies (cf. [2],[42],[79] for example).

Examples F and G appear in GARLING [23]. The mixed topology of Example K has been used to study Banach spaces with bases by SUBRAMANIAN and ROTHMAN (cf. [71],[72]).

<u>I.3.</u> The term "Saks space" was introduced by Orlicz in [42]. We quote his definition (p. 240):

> "Let X be a Banach space or an incomplete Banach space (<u>fundamental space</u>) with the norm $\|\ \|$, and let $\|\ \|$ be another norm defined in X. In the set R of elements $x \in X$ satisfying the inequality $\|x\| \le 1$ we define the distance

between the elements $x,y \in R$ by the formula

$$d(x,y) = ||x - y||^*.$$

If this metric space is complete, it will be termed a Saks space."

We remark that under a norm, Orlicz understands a pseudo-norm or F-norm i.e. the condition of homogeneity is replaced by continuity in the scalar variable. Essentially, this means that the associated topology need not be locally convex. Thus our definition is more general in the sense that the metrisability condition has been dropped, but more restrictive in the sense that local convexity has been demanded. The class of Saks spaces is studied from a categorical point of view by SEMADENI in [65]. Despite this discrepancy between out terminology and that of Orlicz, we have decided to use the term "Saks space" for several reasons. Firstly any reasonable alternative (e.g. "space with mixed topology") is both unwieldy and misleading and secondly the name "Saks space" possesses the same flexibility as the name "Banach space". For example the algebras in the category of Saks space receive automatically the name "Saks algebra".

In 3.12 we have introduced natural notions of tensor product within the category of Saks spaces. This suggests the question: are the associated locally convex spaces the same (algebraically and topologically) as the injective resp. projective tensor product of the spaces, regarded as locally convex spaces with the mixed topologies? The same question has been attacked "from the other end" by NOUREDDINE who showed (in [41]) that if E and and F are semi-Montel D_b-spaces then so is the projective tensor product $E \hat{\otimes} F$. RUESS [59] has shown that the condition of semi-Montelness can be dropped.

The Saks algebras introduced in 3.13 are new. For the theory of Saks spaces, see LABUDA ([33],[34]), LABUDA and ORLICZ [35], ORLICZ ([42],[43],[44],[45]), ORLICZ and PTAK ([46]). The closely related theory of two-norm spaces is developed in ALEXIEWICZ [1] and [2], ALEXIEWICZ and SEMADENI ([7] - [10]), SEMADENI ([63],[64]).

I.4. For 4.1 - 4.3 cf. BRAUNER [14], for 4.4 ALEXIEWICZ and SEMADENI [8], PERSSON [50], WIWEGER [79]. 4.5 is due to WIWEGER [79]. The Banach-Steinhaus problem has been studies intensively - see the papers of ALEXIEWICZ, ORLICZ, ORLICZ and PTAK, LABUDA and LABUDA and ORLICZ. The condition Σ_1 was introduced by ORLICZ in [42] and goes back to the method used by SAKS in [60].

In 4.17 we have adapted the method of partitions of unity for locally convex spaces which was introduced by DE WILDE in [75] (see also KEIM [30]). In 4.23 we have reproduced the proof of a closed graph theorem of KALTON [29]. Note that in the proof of 4.24 we have essentially shown that a locally convex space which has the Banach-Steinhaus property can be used as the source space of a closed graph theorem with the range space a separable

Banach space. In fact, the latter property is equivalent to the Banach-Steinhaus property with separable Banach spaces as range. KALTON has given an internal characterisation of such spaces. 4.29 is a closed graph theorem of BUCHWALTER ([15], 2.4.6).

REFERENCES FOR CHAPTER I.

[1] A. ALEXIEWICZ On sequences of operators II, Studia Math. 11 (1950) 200-236.

[2] On the two norm convergence, Studia Math. 14 (1954) 49-56.

[3] A topology for two-norm spaces, Func. Approx., Comment. Math. 1 (1974) 3-5.

[4] On some two-norm algebras, Func. Approx., Comment. Math. 2 (1976) 3-34.

[5] On some two-norm spaces and algebras, Func. Approx., Comment. Math. 3 (1976) 3-10.

[6] Some two-norm algebras with γ-continuous inverse, Func. Approx., Comment. Math. 3 (1976) 11-21.

[7] A. ALEXIEWICZ, Z. SEMADENI A generalization of two norm spaces. Linear functionals, Bull. Acad. Pol. Sc. 6 (1958) 135-139.

[8] Linear functionals on two norm spaces, Studia Math. 17 (1958) 121-140.

[9] The two norm spaces and their conjugate spaces, Studia Math. 18 (1959) 275-293.

[10] Some properties of two norm spaces and a characterization of reflexivity of Banach spaces, Studia Math. 19 (1960) 116-132.

[11] S. ARIMA Generalized mixed-topologies in dual linear spaces, Yokohama Math. J. 13 (1965) 129-144.

[12] S. ARIMA, M. ORIHARA Generalization of the mixed topology, Yokohama Math. J. 12 (1965) 63-68.

[13] S. BANACH Théorie des opérations linéaires (New York, 1963).

[14] K. BRAUNER Duals of Fréchet spaces and a generalisation of the Banach-Dieudonné theorem, Duke Math. J. 40 (1973) 845-855.

[15] H. BUCHWALTER Espaces vectoriels bornologiques, Publ. Dép. Math. Lyon 2-1 (1965) 2-53.

[16] E. ČECH Topological spaces (Prague, 1966).

[17] G. CHOQUET Sur un theorème du type Banach-Steinhaus pour les convexes topologiques, Sem. Choquet (1973/74) Comm. 4.

[18] H.S. COLLINS Completeness and compactness in linear topological spaces, Trans. Amer. Math. Soc. 79 (1955) 256-280.

[19] C.H. COOK, J. DAZORD Sur la topologie mixte de Wiweger, Publ. Dép. Math. Lyon 11-3 (1974) 1-28.

[20] P. and S. DIEROLF On linear topologies determined by a family of subsets of a topological vector space, to appear in "General Top. and Applications".

[21] R.M. DUDLEY On sequential convergence, Trans. Amer. Math. Soc. 112 (1964) 483-507.

[22] D. van DULST (Weakly) compact mappings into (F)-spaces, Math. Ann. 224 (1976) 111-115.

[23] D.J.H. GARLING A generalized form of inductive limit topology for vector spaces, Proc. London Math. Soc. (3) 14 (1964) 1-28.

[24] I.C. GOHBERG, M.K. ZAMBICKII On the theory of linear operators in spaces with two norms, Transl. Amer. Math. Soc. II 85 (1969) 145-164.

[25] N. de GRANDE-de KIMPE On a class of locally convex spaces with a Schauder basis, Indag. Math. 79 (1976) 307-312.

[26] A. GROTHENDIECK Sur les espaces (F) et (DF), Summa Brasil. Math. 3 (1954) 57-123.

[27] H. HOGBE-NLEND Théorie des bornologies et applications (Springer Lecture Notes 213, 1971).

[28] S.O. IYAHEN, J.O. POPOOLA A generalized inductive limit topology for linear spaces, Glasgow Math. J. 14 (1973) 105-110.

[29] N.J. KALTON Some forms of the closed graph theorem, Proc. Camb. Phil. Soc. 70 (1971) 401-408.

[30] D. KEIM Induktive und projektive Limiten mit Zerlegung der Einheit, Man. Math. 10 (1973) 191-195.

[31] G. KÖTHE Topologische lineare Räume I (Berlin, 1966).

[32] P. KREE Utilisation de limites inductives généralisées d'espaces localement convexes, Sem. Paul Kree (1974/75) Exp. 1.

[33] I. LABUDA Continuity of operators on Saks spaces, Studia Math. 51 (1974) 11-21.

[34] On the existence of non-trivial Saks sets and continuity of linear mappings acting on them, Bull. Acad. Pol. Sc. math., astr., phys., 23 (1975) 885-890.

[35] I. LABUDA, W. ORLICZ Some remarks on Saks spaces, Bull. Acad. Pol. Sc. math., astr., phys. 22 (1974) 909-914.

[36] A. MARQUINA A note on the closed graph theorem (to appear in "Arkiv der Math.").

[37] K. NOUREDDINE Nouvelles classes d'espaces localement convexes, C.R. Acad. Sc. 276 (1973) 1209-1212.

[38] Espaces du type D_b, C.R. Acad. Sc. 276 (1973) 1301-1303.

[39] Nouvelles classes d'espaces localement convexes, Publ. Dép. Math. Lyon 10-3 (1973) 259-277.

[40] Note sur les espaces D_b, Math. Ann. 219 (1976) 97-103.

[41] Localisation topologique, espaces D_b et topologies strictes (Dissertation, Lyon 1977).

[42] W. ORLICZ Linear operations in Saks spaces I, Studia Math. 11 (1950) 237-272.

[43] Linear operations in Saks spaces II, Studia Math. 15 (1955) 1-25.

[44] Contributions to the theory of Saks spaces, Fund. Math. 44 (1957) 270-294.

[45] On the continuity of linear operators in Saks spaces with an application to the theory of summability, Studia Math. 16 (1957) 69-73.

[46] W. ORLICZ, V. PTAK Some remarks on Saks spaces, Studia Math. 16 (1957) 56-68.

[47] B. PERROT Sur la comparaison de certaines topologies mixtes dans les espaces binormés, Colloqu. Math. 34 (1975) 81-90.

[48] Réflexivité dans les espaces mixtes. Application à la réflexivité des espaces bornologiques convexes, C.R. Acad. Sci. 278 (1974) A 1033-1035.

[49] B. PERROT Topologies mixtes dans le cas de structures vectorielles non necessairement convexes, Publ. Dép. Math. Lyon 10-4 (1973) 359-370.

[50] A. PERSSON A generalization of two norm spaces, Ark. f. Math. 5 (1963) 27-36.

[51] H. PFISTER Über eine Art von gemischter Topologie und einen Satz von Grothendieck über (DF)-Räume, Man. Math. 10 (1973) 273-287.

[52] Über das Gewicht und den Überdeckungstyp von uniformen Räumen und einige Formen des Satzes von Banach-Steinhaus, Man. Math. 20 (1977) 51-72.

[53] T. PRECUPANU Remarques sur les topologies mixtes, An. Sti. Univ. "Al. I. CUZA" Iasi Sec. Ia Math. 13 (1967) 277-284.

[54] V. PTAK Completeness and the open mapping theorem, Bull. Soc. Math. France 86 (1958) 41-74.

[55] D. RAIKOV On completeness in locally convex spaces (Russian), Uspehi Mat. Nauk. 14.1 (85) (1959) 223-229.

[56] A. ROBERTSON, W. ROBERTSON Topologische Vektorräume (Mannheim, 1967).

[57] W. ROELCKE On the finest locally convex topology agreeing with a given topology on a sequence of absolutely convex sets, Math. Ann. 198 (1972) 57-80.

[58] On the behaviour of linear mappings on absolutely convex sets and A. Grothendieck's completion of locally convex spaces, III. J. Math. 17 (1973) 311-316.

[59] W. RUESS Halbnorm-Dualität und induktive Limestopologien in der Theorie lokalkonvexer Räume (Habilitationsschrift, Bonn 1976).

[60] S. SAKS — On some functionals, Trans. Amer. Math. Soc. 35 (1933) 549-556.

[61] H. SCHAEFER — Topological vector spaces (New York, 1966).

[62] L. SCHWARTZ — Probabilités cylindriques et applications radonifiantes, J. Fac. Sci. Univ. Tokyo, Sect. I A 18 (1971) 139-186.

[63] Z. SEMADENI — Extensions of linear functionals in two norm spaces, Bull. Pol. Acad. Sci. 8 (1960) 427-432.

[64] Embedding of two norm spaces into the space of bounded continuous functions on a half straight line, Bull. Pol. Acad. Sci. 8 (1960) 421-426.

[65] Projectivity, injectivity and duality, Dissertationes Math. 35 (1963) 1-47.

[66] Banach spaces of continuous functions I, (Warsaw, 1971).

[67] R. SERAFIN — On some class of locally convex spaces connected with Saks spaces and two-norm spaces, Bull. Acad. Polon. Sci. 22 (1974) 1121-1127.

[68] J.H. SHAPIRO — Non-coincidence of the strict and strong operator topologies, Proc. Amer. Math. Soc. 35 (1972) 81-87.

[69] Weak topologies on subspaces of C(S), Trans. Amer. Math. Soc. 157 (1971) 471-479.

[70] K.D. STROYAN — A non standard characterization of mixed topologies (in Springer Lecture Notes 369, 1974).

[71] P.K. SUBRAMANIAN Two-norm spaces and decompositions of Banach spaces, I. Studia Math. 43 (1972) 179-194.

[72] P.K. SUBRAMANIAN, S. ROTHMAN Two-norm spaces and decompositions of Banach spaces, II. Trans. Amer. Math. Soc. 181 (1973) 313-327.

[73] R.F. WHEELER The equicontinuous weak * topology and semi-reflexivity, Studia Math. 41 (1972) 243-256.

[74] M. DE WILDE Limites inductives d'espaces linéaires a seminormes, Bull. Soc. Roy. Sc. Liège 32 (1963) 476-484.

[75] Inductive limits and partitions of unity, Man. Math. 5 (1971) 45-58.

[76] Various types of barrelledness and increasing sequences of balanced and convex sets in locally convex spaces (in:"Summer School on Topological Vector Spaces" Springer Lecture Notes 331, Berlin 1973, pp. 211-217).

[77] M. DE WILDE, C. HOUET On increasing sequences of absolutely convex sets in locally convex spaces, Math. Ann. 192 (1971) 257-261.

[78] A. WIWEGER A topologisation of Saks spaces, Bull. Pol. Acad. Sci. 5 (1957) 773-777.

[79] Linear spaces with mixed topology, Studia Math. 20 (1961) 47-68.

[80] Some applications of the mixed topology to two normed spaces, Bull. Pol. Acad. Sci. 9 (1961) 571-574.

CHAPTER II - SPACES OF BOUNDED, CONTINUOUS FUNCTIONS

Introduction: Chapter II is devoted to the most important and well-developed application of mixed topologies - the theory of strict topologies on spaces of bounded, continuous functions. Since BUCK's original paper (1958) the literature on this topic has grown rapidly. We have tried to give a fairly complete account of this theory from the point of view of mixed topologies. This approach often allows greater generality and simpler and clearer proofs than the original methods.

In section 1, we consider the basic properties of strict topologies. The original strict topology on $C^\infty(X)$ is the mixed topology $\gamma[\|\ \|, \tau_K]$ (see I.2.D) where τ_K is the topology of compact convergence. However, no particular difficulties arise when we replace the topology of compact convergence by that of convergence on an (almost) arbitrary family S of subsets of X. The first part is devoted to relating the elementary properties of the associated mixed topology with properties of X (resp. S). In 1.11 we show how these mixed topologies can be defined by weighted seminorms thus establishing the relation with Buck's topology. In 1.13 we give a very general Stone-Weierstraß theorem based on a result of NEL. We use this to attack the problem of the separability of $C^\infty(X)$. Using partitions of unity we give a new proof of a result of CONWAY-LE CAM.

In § 2 we study the algebraic structure of $C^{\infty}(X)$. We identify its spectrum and show that the topological properties of X are faithfully reflected in the topological-algebraic properties of $C^{\infty}(X)$. We obtain a theory of GELFAND-NAIMARK type and characterise the β-closed ideals of $C^{\infty}(X)$. We then sketch a functional-analytic approach to the real compactification of X.

In § 3 we characterise the dual of $C^{\infty}(X)$ as the space of bounded Radon measures on X, thus extending a result of Buck for locally compact spaces and establishing the basis for applications of strict topologies to measure theory. Tight sets of measures are shown to coincide with the β-equicontinuous sets and the results are used to give a proof of PROHOROV's theorem on the existence of projective limits of compatible families of measures.

The fourth section is devoted to spaces of vector-valued continuous functions. The main results are a tensor-product representation (generalising the classical representation for Banach space valued functions on compact spaces), an exponential law and a tensor-product representation of spaces of functions on product spaces. We also characterise the β-closed $C^{\infty}(X)$-submodules - thus generalising the Stone-Weierstraß theorem.

In the last section we define new strict topologies on $C^{\infty}(X)$ which have the spaces of σ-additive (resp. τ-additive) measures on X as dual. We use the imbedding of X in its Stone-Čech-compactification. We then give some results relating to the problem of establishing a connection between the topological and measure-theoretical properties of X.

II.1. THE STRICT TOPOLOGIES

1.1. In this chapter, X will always denote a completely regular, Hausdorff topological space, $\mathcal{S}$ a saturated family of closed subsets of X (that is, $\bigcup \mathcal{S}$ is dense in X and $\mathcal{S}$ is closed under the formation of finite unions and of closed subsets). Such a family is of countable type if there is a countable subfamily $\mathcal{S}_1$ so that each $B \in \mathcal{S}$ is contained in some $B_1 \in \mathcal{S}$.

Examples of saturated families are:

$\mathcal{F}$: the finite subsets of X;

$\mathcal{K}$: the compact subsets of X;

$\mathcal{B}$: the bounded, closed subsets of X (recall that a subset $B \subseteq X$ is bounded if each continuous, real-valued function on X is bounded on B).

We denote by

C(X) - the space of continuous, complex-valued functions on X;

$C^\infty(X)$ - the space of bounded, continuous, complex-valued functions on X.

$\| \ \|_\infty$ denotes the supremum norm on $C^\infty(X)$. If $B \subseteq X$ then

$$p_B : x \longmapsto \sup \{|x(t)| : t \in B\}$$

is a seminorm on $C^\infty(X)$. If B is bounded, we can regard it as a seminorm on $C(X)$. If S is a saturated family, then τ_S denotes the locally convex topology defined by the seminorms $\{p_B : B \in S\}$. Then $(C^\infty(X), \| \|_\infty, \tau_S)$ is a Saks space and we denote by β_S the associated mixed topology.

1.2. <u>Proposition</u>: 1) $\tau_S \subseteq \beta_S \subseteq \tau_{\| \|_\infty}$ and $\tau_S = \beta_S$ on the norm bounded sets;

2) a subset of $C^\infty(X)$ is β_S-bounded if and only if it is norm-bounded;

3) a sequence (x_n) in $C^\infty(X)$ is β_S-convergent to zero if and only if it is norm bounded and τ_S-convergent to zero;

4) a subset of $C^\infty(X)$ is relatively β_S-compact if and only if it is norm bounded and relatively τ_S-compact;

5) $(C^\infty(X), \beta_S)$ is barrelled or bornological if and only if $\beta_S = \tau_{\| \|_\infty}$;

6) a linear operator T from $C^\infty(X)$ into a locally convex space is β_S-continuous if and only if its restriction to the unit ball of $C^\infty(X)$ is τ_S-continuous;

7) if S is of countable type, then the unit ball of $C^\infty(X)$ is τ_S-metrisable and so a linear mapping T from $C^\infty(X)$ into a locally convex space is β_S-continuous if and only if it is sequentially continuous.

1.3. <u>Corollary</u>: A subset of $C^\infty(X)$ is β_K-precompact if and only if it is norm bounded and equicontinuous.

1.4. <u>Definition</u>: Let S_1 and S_2 be saturated families in X. X is S_1-S_2 <u>normal</u> if for every pair A,B of disjoint, closed subsets of X with $A \in S_1$, $B \in S_2$ there is a continuous function $x : X \longrightarrow [0,1]$ with

$$x|_A = 0 \quad \text{and} \quad x|_B = 1.$$

We say that X is S-<u>normal</u> if X is S-P normal where P is the family of all closed subsets of X.

Note that every X is K-normal (in fact, for a Hausdorff space, K-normality is equivalent to complete regularity - see BUCHWALTER [21], Prop. 2.1.5).

The following Proposition is a generalisation of URYSOHN's theorem and can be proved in exactly the same way (cf. BUCHWALTER [21], Théorème 2.1.6).

1.5. <u>Proposition</u>: X is S - S normal if and only if for each $A \in S$ and each continuous mapping $x : A \longrightarrow [0,1]$ there is a continuous $\tilde{x} : X \longrightarrow [0,1]$ so that $\tilde{x}|_A = x$.

1.6. <u>Proposition</u>: If X is S - S normal then $\beta_S = \tau_{\|\ \|_\infty}$ if and only if $X \in S$.

<u>Proof</u>: The sufficiency of the condition $S \in B$ is trivial. Suppose that $X \notin S$. Then we can find, for each seminorm p_B $(B \in S)$, an x in the unit ball of $C^\infty(X)$ so that $p_B(x) = 0$ and $\|x\|_\infty = 1$.

1.7. In the following Proposition, we examine the problem of the completeness of $(C^\infty(X),\beta_S)$. It is convenient to introduce the following concept: a space X is S-complete if each mapping $x : X \longrightarrow \mathbb{C}$ is continuous if and only if $x|_A$ is continuous for each $A \in S$ (it suffices to consider bounded functions). For example, X is F-complete if and only if it is discrete. The K-complete spaces are precisely the k_R-spaces (see, for example, MICHAEL [100]).
Locally compact spaces and metrisable spaces are K-complete. To each space X one can associate a S-complete space in a natural way: we give X the weak topology defined by the family of mappings from X into $\mathbb{C}$ which are such that their restrictions to each $A \in S$ are continuous. We denote X with this topology by X_S. Then X_S is S-complete and X is S-complete if and only if $X = X_S$. $C^\infty(X_S)$ is precisely the space of bounded mappings from X into $\mathbb{C}$ whose restrictions to the sets of S are continuous.

1.8. Proposition: Suppose that $\cup S = X$. Then $(C^\infty(X),S)$ is complete if X is S-complete. The converse is true if X is S - S normal.

Proof: Suppose that X is S-complete. Let $(x_\alpha)_{\alpha\in I}$ be a τ_S-Cauchy net in $B_{\|\,\|_\infty}$, the unit ball of $C^\infty(X)$. Then (x_α) converges pointwise to a function x from X into $\mathbb{C}$ and the restriction of x to $A \in S$ is continuous, as the uniform limit of $(x_\alpha|_A)_{\alpha\in I}$. Hence x is continuous and $x_\alpha \longrightarrow x$ in β_S.

Now suppose that $(C^{\infty}(X),\beta_{S})$ is complete and that $x : X \longrightarrow \mathbb{C}$ is such that $x|_{A}$ is continuous for each $A \in S$. We show that x is continuous when X is $S-S$ normal. It is no loss of generality to suppose that $x : X \longrightarrow [0,1]$. For each $A \in S$, let x_A be a continuous function from X into [0,1] so that $x_A = x$ on A (1.5). Then $(x_A)_{A\in S}$ is a Cauchy net for β_S and so converges to a function in $C^{\infty}(X)$ i.e. $x \in C^{\infty}(X)$.

1.9. <u>Corollary</u>: 1) $(C^{\infty}(X),\beta_{K})$ is complete if and only if X is K-complete;
2) $(C^{\infty}(X),\beta_{F})$ is complete if and only if X is discrete.

1.10. <u>Corollary</u>: Suppose that X is $S-S$ normal and that $\cup S = X$. Then the completion of $(C^{\infty}(X),\beta_{S})$ is $(C^{\infty}(X_{S}),\beta_{S})$.

These results can be interpreted as follows: under the restriction operators, the family $\{C^{\infty}(A); A \in S\}$ of Banach spaces forms a projective spectrum. If $X = \cup S$ then the Saks space projective limit of this spectrum is naturally identifiable with $(C^{\infty}(X_S),\|\ \|_{\infty},\tau_S)$, in particular, with $(C^{\infty}(X),\|\ \|_{\infty},\tau_S)$ if X is S-complete.

Now we give a concrete representation of a family of seminorms which defines β_S. From these one can easily deduce that the mixed topology β_S reduces, in special cases, to the strict topologies which have been studied on spaces of bounded, continuous functions. We denote by L_S^+ the set of bounded non-negative upper semi-continuous functions ϕ on X which <u>vanish at infinity with respect to</u> S i.e. which satisfy the condition:

$$\text{for each } \varepsilon > 0, \quad \{t \in S : \phi(t) \geq \varepsilon\} \in S$$

If $\phi \in L_S^+$,

$$p_\phi : x \longmapsto \|\phi x\|_\infty$$

is a seminorm on $C^\infty(X)$. The family of all such seminorms defines a locally convex topology $\tilde{\beta}$ on $C^\infty(X)$ (note that the characteristic function of each $A \in S$ is in L_S^+ and so $\tilde{\beta}$ is finer than τ_S).

1.11. Proposition: If X is $S - S$ normal, then $\tilde{\beta} = \beta_S$.

Proof: We note first that $(C^\infty(X), \| \|_\infty, \tau_S)$ satisfies condition (a) of I.4.5. If $A \in S$ and $x \in C^\infty(X)$ then, by 1.5, there is a $y \in C^\infty(X)$ so that $y = x$ on A and $\|y\|_\infty = p_A(x)$. Let $z : x - y$. Then $p_A(z) = 0$ and so $x = y + z$ is a suitable decomposition of x.

We now show that β_S is finer than $\tilde{\beta}$ i.e. that $\tilde{\beta}$ is coarser than τ_S on $B_{\| \|_\infty}$. If $\phi \in L_S^+$, $\varepsilon > 0$ and $A := \{t \in S : \phi(t) \geq \varepsilon\}$ then for $x \in B_{\| \|_\infty}$ with $p_A(x) \leq \{\varepsilon \sup_{t \in X} |\phi(t)|\}^{-1}$ we have $p_\phi(x) \leq \varepsilon$. On the other hand, if V is a β-neighbourhood of zero, then, by I.4.5, it contains a set of the form

$$\{x \in C^\infty(X) : p_{A_n}(x) \leq \lambda_n\}$$

where (A_n) is an increasing sequence in S and (λ_n) is a strictly increasing sequence of positive numbers which converges to infinity. Then if

$$\phi : t \longmapsto \begin{cases} \lambda_1^{-1} & (t \in A_1) \\ \lambda_n^{-1} & (t \in A_n \setminus A_{n-1}) \\ 0 & (t \in X \setminus \cup A_n) \end{cases}$$

ϕ is in L_S^+ and V contains the unit ball of p_ϕ.

We now give a Stone-Weierstraß theorem for β_S. For convenience, we consider the space $C_{\mathbb{R}}^\infty(X)$ of real valued functions on X as a real vector space. The results can be extended to complex-valued functions using standard methods. Since the sets of S are not necessarily compact, we need a refinement of the classical Stone-Weierstraß theorem due to NEL. Recall that a subset of X is a <u>zero-set</u> if it has the form $x^{-1}(0)$ for some $x \in C_{\mathbb{R}}^\infty(X)$. A subset M of $C_{\mathbb{R}}^\infty(X)$ <u>separates disjoint zero-sets</u> if for each pair A,B of disjoint zero sets in X, there is an $x \in M$ so that $\overline{x(A)}$ and $\overline{x(B)}$ are disjoint. The following Lemma follows from the fact that the points of the Stone-Čech compactification βX of X are limits of z-ultrafilters in X (cf. GILLMAN and JERISON [55], Ch. 6). Its proof can be found in NEL [108].

1.12. <u>Lemma</u>: Let M be a subset of $C_{\mathbb{R}}^\infty(X)$ which separates disjoint zero sets in X. Then M, regarded as a subset of $C_{\mathbb{R}}^\infty(\beta X)$, separates the points of βX.

1.13. <u>Proposition</u>: Suppose that M is a subalgebra of $C_{\mathbb{R}}^\infty(X)$ so that for each $A \in S$, M_A, the restriction of M to A, separates disjoint zero-sets in A and contains a function which is bounded away from zero on A. Then M is β_S-dense in $C^\infty_{\mathbb{R}}(X)$.

Proof: We can assume that M is β_S-closed. Then it is norm closed and so is a lattice under the pointwise ordering. We show that if $x \in C^\infty_{\mathbb{R}}(X)$, $0 < \varepsilon < 1$ and if $A \in S$, then there is a $y \in M$ so that $\|y\|_\infty \leq \|x\|_\infty + 1$ and $p_A(x-y) \leq \varepsilon$. M_A, regarded as an algebra of functions on βA, satisfies the conditions of the classical Stone-Weierstraß theorem (for M_A seperates the points of βA by 1.12). Hence M_A is norm-dense in $C^\infty_{\mathbb{R}}(A)$ and so there is a $y_1 \in M$ with $p_A(x-y_1) \leq \varepsilon$. Then

$$y := \sup\{\inf(y_1, \|x\|_\infty + 1), -(\|x\|_\infty + 1)\}$$

is the required function.

1.14. Corollary: Let M be a subalgebra of $C^\infty_{\mathbb{R}}(X)$ so that M separates the points of X and for each $t \in X$ there is an $x \in M$ with $x(t) \neq 0$. Then M is β_K-dense in $C^\infty_{\mathbb{R}}(X)$.

We now consider the problem of characterising those spaces for which $(C^\infty(X), \beta_S)$ is separable. Since the Banach space $C^\infty(A)$ $(A \in S)$ can only be separable if A is compact (this is a classical result of M. KREIN and S. KREIN and follows from the fact that the Stone-Čech compactification of a non-compact space is never metrisable - see GILLMAN and JERISON [55], § 9.6) and $C^\infty(A)$ is, at least when X is $S-S$ normal, a continuous image of $(C^\infty(X), \beta_S)$, it is natural to impose the condition that $S \subseteq K$ i.e. the sets of S are compact.

1.15. <u>Proposition</u>: If $\mathcal{F} \subseteq \mathcal{S} \subseteq \mathcal{K}$ then $(C^{\infty}(X), \beta_{\mathcal{S}})$ is separable if and only if there is a weaker separable, metrisable topology on X.

<u>Proof</u>: Since we shall use the Stone-Weierstraß theorem, it is convenient to restrict attention to $C^{\infty}_{\mathbb{R}}(X)$.
If M is a countable $\beta_{\mathcal{S}}$-dense subset of $C^{\infty}_{\mathbb{R}}(X)$, then the weak topology defined by M satisfies the given conditions.
Sufficiency: let τ be a suitable separable, metrisable topology on X. Then by Urysohn's metrisation theorem (WILLARD [157], 23.1) (X,τ) can be embedded in a compact, metrisable space Y. For each positive integer n, we can find a finite open covering $(\mathcal{U}_n)$ of Y so that $\max\{\operatorname{diam} U : U \in \mathcal{U}_n\} \leq 1/n$ (diam (U) is the diameter of U). Let Φ_n be a partition of unity of Y subordinate to $\mathcal{U}_n$ and denote by M the subalgebra of $C^{\infty}_{\mathbb{R}}(X)$ generated by the restrictions of the elements of $\bigcup_n \Phi_n$ to X. Then M is $\beta_{\mathcal{S}}$-dense by 1.14 and so $(C^{\infty}_{\mathbb{R}}(X), \beta_{\mathcal{S}})$ is separable.

1.16. <u>Remark</u>: It follows from SMIRNOV's metrisation theorem (see WILLARD [157], 23.G.3) that if X is locally compact and paracompact and S possesses a weaker metrisable topology, then X is metrisable. Hence if X is locally compact, paracompact and $(C^{\infty}(X), \beta_{\mathcal{S}})$ is separable, then X is metrisable (cf. SUMMERS [139], Th. 2.5).

1.17. Proposition: If X is discrete, then $(C^\infty(X),\beta_S)$ is separable if and only if card (X) $\leq$ card ($\mathbb{R}$) .

Proof: The necessity follows from 1.15 and the fact that the cardinality of a separable metrisable space is at most card ($\mathbb{R}$). On the other hand, $\mathbb{R}$ (and hence any subset) has a separable, metrisable topology - the natural topology.

1.18. Remark: A more intricate argument shows that the same result holds for metrisable spaces X (see SUMMERS [139], Th. 3.2)

In the third section of this chapter we shall consider duality for $C^\infty(X)$, with strict topologies. However, using the theory of Chapter I we can already provide some information on this duality, without specifically calculating the dual space - in particular, we can give sufficient conditions for β_K to be the Mackey topology and we can characterise the relatively weakly compact subsets of $C^\infty(X)$.

1.19. Partitions of unity for $C^\infty(X)$: Let X be a locally compact, paracompact space. Then there exists a partition $\{\phi_K : K \in \mathcal{K}\}$ of unity on X (see BOURBAKI [15], IX.4.3) so that supp $\phi_K \subseteq K$. Now $(C^\infty(X), \|\ \|, \tau_{\mathcal{K}})$ is the Saks space projective limit of the system $\{C(K) : K \in \mathcal{K}\}$ of Banach spaces and if we define the mappings

$$T_K : x \longmapsto (x\,\phi_K)^\wedge$$

from C(K) into $C^{\infty}(X)$ where $(x\,\phi_K)^{\wedge}$ denotes the extension of $x\,\phi_K$ to a function on X obtained by setting it equal to zero off K, then $\{T_K\}$ is a partition of unity in the sense of I.4.17. Hence we have, by I.4.19 and I.4.24:

1.20. Proposition: Let X be locally compact and paracompact. Then $(C^{\infty}(X), \beta_K)$ is a Mackey space and has the Banach-Steinhaus property. Also a linear mapping from $C^{\infty}(X)$ into a separable Fréchet space is β_K-continuous if and only if its graph is closed.

In the next results, the phrase "weak topology on $C^{\infty}(X)$" will be used to denote the weak topology defined by the dual of $(C^{\infty}(X), \beta_K)$.

1.21. Proposition: A sequence (x_n) in $C^{\infty}(X)$ converges weakly to x if and only if $\{x_n\}$ is uniformly bounded and the functions x_n converge pointwise to x.

1.22. Proposition: A bounded subset B of $C^{\infty}(X)$ is weakly precompact if and only if it is precompact for the topology of pointwise convergence on X. Hence if X is K-complete, then B is relatively weakly compact if and only if it is relatively compact for the topology of pointwise convergence on X.

Proof: Using I.1.20, we can reduce 1.21 and 1.22 to the case where X is compact (see, for example, GROTHENDIECK [62], pp. 12 and 209 for this case).

1.23. Remark: Using results from GROTHENDIECK [60], one can strengthen 1.22 as follows: suppose that X is $\mathcal{K}$-complete and has a dense subset which is the union of countably many compact sets. Then the following conditions on a bounded subset B of $C^\infty(X)$ are equivalent:

a) B is relatively countable compact for τ_p (resp. for the weak topology);

b) B is relatively sequentially compact for τ_p (resp. for the weak topology);

c) B is relatively compact for τ_p (resp. for the weak topology).

Here τ_p denotes the topology of pointwise convergence on X.

1.24. Remark: One of the main themes of this Chapter will be that of relating the topological properties of X with the linear (or algebraic) and topological properties of $C^\infty(X)$ with various mixed topologies (cf. 1.8, 1.9, 1.15 for example). We list here some examples without proofs:

1) X is hemi-compact (i.e. $\mathcal{K}$ is of countable type) if and only if $B_{\|\ \|_\infty}$ is $\beta_{\mathcal{K}}$-metrisable;

2) $B_{\|\ \|_\infty}$ is $\beta_{\mathcal{K}}$-separable and metrisable if and only if X is hemi-compact and each $K \in \mathcal{K}$ is metrisable;

3) if X is locally compact, then $B_{\|\ \|_\infty}$ is $\beta_{\mathcal{K}}$-separable and metrisable if and only if X is separable and metrisable (alternatively if X is the countable union of compact, metrisable sets);

4) the following conditions are equivalent:

a) $B_{\| \|}$ is β_K-compact (i.e. $(C^\infty(X),\beta_K)$ is semi-Montel);

b) $(C^\infty(X),\beta_K)$ is semi-reflexive;

c) $(C^\infty(X),\beta_K)$ is a Schwartz space;

d) X is discrete.

5) $(C^\infty(X),\beta_K)$ is nuclear if and only if X is finite;

6) $\beta_K = \tau_K$ on $C^\infty(X)$ if and only if the union of countably many compact subsets of X is relatively compact. If this is the case, then $(C^\infty(X),\beta_K)$ is a (DF)-space.

1.25. Remark: A number of results given in this section for $C^\infty(X)$ with the topology β_K (e.g. those of 1.24) can be extended to β_S with the natural changes. We leave the task of carrying out such extension to the interested reader, mentioning only that 1.20 can be extended to the topology $\beta_{\mathcal{B}}$ by replacing the assumption of local compactness by local boundedness (obvious definition !) and that SCHMETS and ZAFARANI have studied the topology β_P in [124].

II.2. ALGEBRAS OF BOUNDED, CONTINUOUS FUNCTIONS

In the first part of this section, we work exclusively with the strict topology defined by the family K of compact subsets of X. To simplify the notation, we denote it by β. First we note that $(C^\infty(X),\| \|,\tau_K)$ is a pre-Saks algebra (that is, its completion is a Saks algebra).

2.1. Proposition: Multiplication is continuous on $(C^\infty(X),\beta)$.

Proof: We use the representation of β given in 1.11. If $\phi \in L_K^+$, so does $\psi := \sqrt{\phi}$ and we have the following inequality

$$p_\phi(xy) \leq p_\psi(x)p_\psi(y) \qquad (x,y \in C^\infty(X)).$$

In general, inversion is not continuous on $C^\infty(X)$ and so $(C^\infty(X),\beta)$ is not a locally multiplicatively convex algebra in the sense of MICHAEL [99].

If $t \in X$ then

$$\delta_t : x \longmapsto x(t)$$

is an element of the spectrum $M_\gamma(C^\infty(X))$ of $C^\infty(X)$. We have thus constructed a mapping $\delta : t \longrightarrow \delta_t$ from X into $M_\gamma(C^\infty(X))$. We call it the generalised Dirac transformation. It is injective since $C^\infty(X)$ separates X.

2.2. Proposition: δ is a homeomorphism from X onto $M_\gamma(C^\infty(X))$.

Proof: Since the topologies on X and $M_\gamma(C^\infty(X))$ are the weak topologies defined by $C^\infty(X)$, it is sufficient to show that δ is surjective. Let f be a β-continuous multiplicative functional on $C^\infty(X)$ and denote by M the kernel of the restriction of f to $C^\infty_{\mathbb{R}}(X)$. Then there is a $t_o \in X$ so that $x(t_o) = 0$ for each $x \in M$ (for otherwise M would satisfy the conditions of 1.14 and so would be β-dense in $C^\infty_{\mathbb{R}}(X)$ i.e. f would be zero).

Note that M separates X. For otherwise there would be points s_1, s_2 in X so that $x(s_1) = x(s_2)$ for $x \in M$. Then M would lie in the kernel of the linear form $x \longmapsto x(s_1) - x(s_2)$ and so would have codimension at least two. Then $M = \{x : x(t_o) = 0\}$ (for both these sets have codimension one) and so $f = \delta_{t_o}$.

If X, X_1 are completely regular spaces, $\phi : X \longrightarrow X_1$ continuous, then

$$C^\infty(\phi) : x \longmapsto x \circ \phi$$

is a β-continuous star homomorphism from $C^\infty(X_1)$ into $C^\infty(X)$. In fact, every such homomorphism has this form as the following result shows:

2.3. Proposition: If Φ is a β-continuous homomorphism from $C^\infty(X_1)$ into $C^\infty(X)$ then Φ has the form $C^\infty(\phi)$ for some continuous mapping ϕ from X into X_1.

Proof: If $t \in X$ then $\delta_t \circ \Phi$ is a β-continuous multiplicative form on $C^\infty(X_1)$ and so is defined by a unique element of X_1 - we denote this element by $\phi(t)$. By the construction of this mapping ϕ we have

$$\Phi(x) = x \circ \phi$$

for each $x \in C^\infty(X_1)$. Hence for each $x \in C^\infty(X_1)$, $x \circ \phi \in C^\infty(X)$ and this property characterises continuity for mappings between completely regular spaces.

2.4. <u>Corollary</u>: X and X_1 are homeomorphic if and only if $C^\infty(X)$ and $C^\infty(X_1)$ are isomorphic as pre-Saks algebras.

We remark that the following version of the Banach-Stone theorem for non-compact spaces can be deduced from 2.4: if there is an isometry from $C^\infty(X)$ onto $C^\infty(X_1)$ which is also β-bicontinuous, then X and X_1 are homeomorphic.

If $(A, \|\ \|, \tau)$ is a commutative pre-Saks algebra with unit and if $x \in A$, then the mapping

$$\hat{x} : f \longrightarrow f(x)$$

from $M_\gamma(A)$ into $\mathbb{C}$ is an element of $C^\infty(M_\gamma(A))$. Thus we have constructed an algebra homomorphism from A into $C^\infty(M_\gamma(A))$. We call it the <u>generalised Gelfand-Naimark transform</u>. Note that we can regard $M_\gamma(A)$ as a subspace of the spectrum M(A) of the normed algebra $(A, \|\ \|)$. The generalised Gelfand-Naimark transform is then the composition of the Gelfand-Naimark transform for A and the restriction operator from C(M(A)) into $C^\infty(M_\gamma(A))$. In particular, if A is a pre-Saks C*-algebra, then the generalised Gelfand-Naimark transform is a star-homomorphism (this also follows directly from 2.3).

2.5. <u>Proposition</u>: If $(A, \|\ \|, \tau)$ is a commutative Saks C*-algebra then the generalised Gelfand-Naimark transform is an algebra isomorphism from A onto $C^\infty(M_\gamma(A))$.

Proof: We first note that the image of A in $C^\infty(M_\gamma(A))$ is a self-adjoint, separating subalgebra which contains the constants and so is β-dense by the complex version of 1.14. Now let P be a family of C^*-seminorms on A which define τ (as in I.3.13). For each $p \in P$, we denote by A_p the associated C^*-algebra and by $M(A_p)$ its spectrum. We can regard $M(A_p)$ as a (compact) subset of $M_\gamma(A)$ and we show that $M_\gamma(A) = \bigcup_{p\in P} M(A_p)$. If $f \in M_\gamma(A)$, then, by I.3.10, there is an increasing sequence (p_n) in P and an $f_n \in A_{p_n}'$ so that Σf_n is absolutely summable to f. We can also suppose that $\|\Sigma f_n\| < 1 + \varepsilon$ for an arbitrary positive ε. Choose n_o so that $\sum_{n>n_o} \|f_n\| < \varepsilon$. Then $f \in M(A_{p_{n_o}})$ for small enough ε. For if $f \notin M(A_{p_{n_o}})$ then there is an $x \in C^\infty(M_\gamma(A))$ so that $\|x\| \leq 1$, $x(f) = 1$ and $x = o$ on $M(A_{p_{n_o}})$. By the β-density of the image of A, there is an $x_1 \in A$ with $\|x_1\| \leq 1 + \varepsilon$ and $|\hat{x}_1(f)| \leq 1 - \varepsilon$, $|\hat{x}_1| < \varepsilon$ on $M(A_{p_{n_o}})$. Hence, for each $g \in A_{p_{n_o}}'$ with $\|g\| < 1 + \varepsilon$ we have

$$\|f - g\| \geq (1+\varepsilon)^{-1} \|f(x_1) - g(x_1)\| \geq \frac{1-\varepsilon}{1+\varepsilon} - \varepsilon$$

and we obtain a contradiction for small ε by taking $g = \sum_{n=1}^{n_o} f_n$.

To complete the proof, we let S be the family of closed subsets of $M_\gamma(A)$ which are contained in some $M(A_p)$ (the p depending on the subset) and, as a temporary notation, $\hat{A}$ be the Saks space projective limit of the system $\{C(M(A_p))\}_{p\in P}$.

Consider the following diagram

$$\begin{array}{ccccc}
A & \longrightarrow & A_q & \longrightarrow & A_p \\
\downarrow & & \downarrow & & \downarrow \\
\hat{A} & \longrightarrow & C(M(A_q)) & \longrightarrow & C(M(A_p))
\end{array}$$

where the vertical arrows are the corresponding Gelfand-Naimark transforms and so are isomorphisms. Then the general Gelfand-Naimark transform, being the unique arrow from A into $\hat{A}$ which preserves commutativity, is an isomorphism and so $\hat{A} = C^{\infty}(M_{\gamma}(A))$ and the generalised Gelfand-Naimark transform is surjective.

Note that the inverse of the generalised Gelfand-Naimark transform is β-continuous. However, we cannot, in general, expect it to be bi-continuous. For example, if S is a proper subfamily of K which contains F and is such that a function $x : X \longrightarrow \mathbb{C}$ is continuous if and only if its restriction to the sets of S are continuous, then the generalised Gelfand-Naimark transform for $(C^{\infty}(X), \| \|, \tau_S)$ is (up to the obvious identifications) the identity from $(C^{\infty}(X), \| \|, \tau_S)$ into $(C^{\infty}(X), \| \|, \tau_K)$ and this is not continuous, in general (as an example of such an S we could take the family consisting of the ranges of convergent sequences and their limit points in a metrisable space).

We now characterise local compactness for X in terms of properties of $C^{\infty}(X)$. Let $(A, \| \|, \tau)$ be a commutative Saks algebra and let

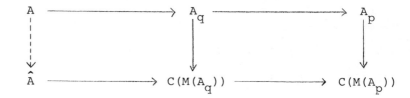

P be a suitable family of submultiplicative seminorms defining τ. If $p \in P$, put

$$I_p := \{x \in A : p(x) = 0\}$$
$$A(I_p) := \{y \in A : yI_p = 0\}.$$

A is <u>perfect</u> if $\sum_{p \in P} A(I_p)$ is γ-dense in A. Obviously this property is preserved if we refine the topology τ. As an example, if p is the seminorm p_K $(K \in \mathcal{K})$ on $C^\infty(X)$, then

$$A(I_p) = \{x \in C^\infty(X) : x(t) = 0 \quad \text{for} \quad t \in X \setminus K\}$$

Hence $A(I_p)$ is $C_c(X)$, the space of functions in $C^\infty(X)$ with compact support.

2.6. <u>Proposition</u>: A completely regular space is locally compact if and only if $(C^\infty(X), \|\ \|, \tau_{\mathcal{K}})$ is perfect.

<u>Proof</u>: In view of the above remarks, this is equivalent to the following statement: X is locally compact if and only if $C_c(X)$ is β-dense in $C^\infty(X)$.
Suppose that X is locally compact. Then $C_c(X)$ separates X and so is β-dense by 1.14.
Now suppose that $C_c(X)$ is β-dense in $C^\infty(X)$. If $t \in X$, then there is an $x \in C_c(X)$ so that $x(t) > 0$. Then

$$\{s : x(s) > 0\}$$

is a relatively compact neighbourhood of t.

Using the generalised Gelfand-Naimark transform, it is easy to see that if A is a perfect, commutative Saks C*-algebra, then $M_\gamma(A)$ is locally compact. The reverse implication is not true.

Let I be an ideal in $C^\infty(X)$ and write

$$Z(I) = \bigcap_{x \in I} Z(x)$$

where $Z(x) := x^{-1}(0)$ is the zero-set of x. Then we put

$$I(Z(I)) := \{y \in C^\infty(X) : y = 0 \text{ on } Z(I)\}.$$

Then $I(Z(I))$ is obviously a β-closed ideal and $I \subseteq I(Z(I))$. We shall now show that $I = I(Z(I))$ if and only if I is β-closed. This result is well-known for compact X and we shall use it for the proof of the result in the general case. We sketch briefly how it can be proved. Suppose that $x \in I(Z(I))$. For each $\varepsilon > 0$ we can find an open neighbourhood U of $Z(I)$ and a function x_ε in $C(X)$ so that x_ε vanishes on U and $\|x - x_\varepsilon\| \le \varepsilon$. We shall show that $x_\varepsilon \in I$ which will finish the proof. By a compactness argument, there exist $x_1,\dots,x_n$ in I so that $U \supseteq \bigcap_{i=1}^{n} Z(x_i)$. Then U contains the zero-set of the element $y := \sum_i |x_i|^2$ of I. But then the zero-set of x_ε is a neighbourhood of the zero-set of y and so x_ε is a multiple of y.

2.7. Proposition: Let I be a β-closed ideal of $C^\infty(X)$. Then

$$I = I(Z(I)).$$

Proof: I is a norm-closed ideal in $C(\beta X)$ and so there is a closed set K_0 in βX so that $I = \{x \in C(\beta X) : x = 0 \text{ on } K_0\}$.

It is obviously sufficient to show that $K_o = cl_{\beta X} Z(I)$ (closure in βX) for then if a function vanishes on $Z(I)$ its extension to βX vanishes on K_o and so is in I. If this were not the case, there would be a $t_o \in K_o \setminus cl_{\beta X} Z(I)$. Then there is a $y_o \in C(\beta X)$ with $y_o(t_o) = 1$ and $y = 0$ on a neighbourhood of $cl_{\beta X} Z(I)$.

We now show that $y_o \in I$ which gives a contradiction. To do this, we show that for each $K \in \mathcal{K}$, $\varepsilon > 0$, there is a $y_{K,\varepsilon}$ in I so that $p_K(y_o - y_{K,\varepsilon}) \leq \varepsilon$ and $\|y_{K,\varepsilon}\| \leq \|y_o\|$. Then $(y_{K,\varepsilon})$ is a net in I which is $\beta_{\mathcal{K}}$-convergent to y_o. To construct $y_{K,\varepsilon}$ we proceed as follows: let I_K denote the projection of I in $C(K)$. Then I_K is an ideal in $C(K)$ and so $\overline{I}_K$, its closure in the Banach space $C(K)$, is a closed ideal in $C(K)$. Hence it has the form $\{y \in C(K) : y = 0 \text{ on } Z(I) \cap K\}$. Hence $y|_K \in \overline{I}_K$ and so Tietze's theorem implies the existence of the required $y_{K,\varepsilon}$.

The results of this section can be used to give a natural construction of the Stone-Čech compactification and the real-compactification of a completely regular space. We describe this briefly, firstly to display the connection between mixed topologies and the theory of topological extensions and secondly because it will allow us to give a significant generalisation of 2.2.

If X is a completely regular space, $(C^\infty(X), \| \ \|)$ is a Banach algebra. Its spectrum $M(C^\infty(X))$ is a compact space which we denote by βX. The Dirac transformation can be regarded as a (topological) embedding of X into βX. It has the following universal property: if ϕ is a continuous mapping from X into a compact space K, then there is a unique continuous extension $\tilde{\phi}$ of ϕ to a continuous mapping from βX into K. For consider the operator

$$C^{\infty}(\phi) : C(K) \longrightarrow C^{\infty}(X) = C(\beta X)$$

which is $\| \,\|$-β continuous and so (by the closed graph theorem or, more elementarily, by I.1.11) $\| \,\| - \| \,\|$ continuous. Hence, by 2.3, $C^{\infty}(\phi)$ (regarded as a mapping from C(K) into $C(\beta X)$, has the form $C^{\infty}(\tilde{\phi})$ for some $\tilde{\phi} : \beta X \longrightarrow K$. $\tilde{\phi}$ has the required property.

Now we denote by I(X) the set of those functions in $C^{\infty}(X)$ which have no zeros in X (i.e. are invertible in the algebra C(X)). Every $x \in C^{\infty}(X)$ has a unique extension to a function in $C(\beta X)$ which we shall continue to denote by x. Then we put

$$\upsilon X := \bigcap_{x \in I(X)} C_{\beta X}(x)$$

where $C_{\beta X}(x) = \{s \in \beta X : x(s) \neq 0\}$ is the co-zero set of x in βX.

υX is the <u>real-compactification</u> of X and X is <u>real-compact</u> if $\upsilon X = X$. The above rather unfamiliar definition is the natural one from the point of view of strict topologies.

The following equivalent forms are better known:

2.8. Proposition: 1) $\upsilon X = \bigcap_{x \in C_{\mathbb{R}}(X)} \widetilde{x}^{-1}(\mathbb{R})$

where $\widetilde{x}$ denotes the extension of $x \in C_{\mathbb{R}}(X)$ to a function from βX into the 2-point compactification of $\mathbb{R}$.

2) υX is the completion of X with respect to the C(X)-uniformity on X.

Proof: 1) follows from the simple fact that if $x \in I(X)$, then $1/|x| \in C_{\mathbb{R}}(X)$ and the zeros of x in βX are precisely the points where $(1/|x|)^{\sim}$ is infinite in value.
For 2) see GILLMAN and JERISON [55], § 15.13.

2.9. Corollary: For a completely regular space X, the following are equivalent:

1) X is complete for the C(X)-uniformity;
2) X is real-compact;
3) for each $s \in \beta X \setminus X$, there is an $x \in I(X)$ with $x(s) = 0$.

Now it is clear that the bounded subsets of X are precisely those subsets of X which are precompact in the C(X)-uniformity or, by the above result, relatively compact in υX.
This remark makes it natural to generalise the definitions of 1.1 to include saturated families S of subsets of υX (for reasons which will be clear later, it is convenient to drop the assumption that the sets be closed). Hence if S is such a family, we can define the strict topology β_S on $C^\infty(X)$. We shall always assume

that the subsets of S are relatively compact in υX.
The following result is a significant generalisation of 2.2.

2.10. <u>Proposition</u>: The spectrum of the topological algebra $(C^\infty(X),\beta_S)$ is $\bigcup_{B\in S} cl_{\upsilon X}(B)$ (closure in υX).

<u>Proof</u>: It is clear that any point in $\bigcup_{B\in S} cl_{\upsilon X}(B)$ defines an element in the required spectrum. On the other hand, any point in the spectrum is defined by a member of υX (apply 2.2 to the space υX). Hence it is sufficient to show that if $t_o \notin \bigcup_{B\in S} cl_{\upsilon X}(B)$ then δ_{t_o} is not β_S-continuous. But this follows easily from the fact that for each $B \in S$ there is an $x \in B_{\|\,\|_\infty}$ so that $x(t_o) = 1$ and $x = 0$ on B.

2.11. <u>Corollary</u>: The spectrum of $(C^\infty(X),\beta_{\mathcal{B}})$ is $\bigcup_{B\in\mathcal{B}} cl_{\upsilon X}(B)$.

2.12. <u>Remark</u>: The space of 2.11 has been introduced by BUCHWALTER [22] who denoted it by X" (because of a certain formal analogy with the bidual of a locally convex space) in connection with the concept of a μ-<u>space</u> i.e. a completely regular space in which $\mathcal{B} = \mathcal{K}$ (cf. the concept of semi-Montel locally convex space). Every space has a "μ-ification" μX which is obtained as the limit of the transfinite series X,X",(X")",... and X is a μ-space if and only if $X = \mu X$ (or alternatively if X = X"). It is now clear that X is a μ-space if and only if $\beta_{\mathcal{K}} = \beta_{\mathcal{B}}$ on $C^\infty(X)$.

II.3. DUALITY THEORY

A classical result of BUCK for the space $C^{\infty}(X)$ (X locally compact) is that the dual of $(C^{\infty}(X),\beta)$ is the space of bounded Radon measures on X. In this section, we shall extend this result to completely regular spaces. We shall take this opportunity to discuss various equivalent definitions of Radon measures on completely regular spaces.

3.1. Definition: A premeasure on X is a member of the (vector space) projective limit of the system

$$\{\tilde{\rho}_{K_1,K} : M(K_1) \longrightarrow M(K);\ K \subseteq K_1,\ K,K_1 \in \mathcal{K}(X)\}$$

In other words, a premeasure is a system $\mu = \{\mu_K\}$ of Radon measures which satisfies the compatibility relations $\mu_{K_1|K} = \mu_K$ $(K \subseteq K_1)$. If $\mu = \{\mu_K\}$ is a premeasure on X, $|\mu_K|^*$ denotes the outer measure on K defined by $|\mu_K|$ (see BOURBAKI [13] § IV.1.4) Thus $|\mu_K|^*$ is defined as follows: if $U \subseteq K$ is open, $|\mu_K|^*(U)$ is defined to be $\sup\{\int f\, d|\mu_K|\}$ where f ranges over the family of positive, continuous functions on K with $f \le \chi_U$. For general $A \subseteq K$, $|\mu_K|^*(A)$ is defined to be

$$\inf\{|\mu_K|^*(U) : U \text{ open in } K,\ A \subseteq U\}.$$

Now if C is a subset of X we define $|\mu|^*(C)$ to be $\sup\{|\mu_K|^*(C \cap K) : K \in \mathcal{K}(X)\}$.

A premeasure μ on X is said to be tight if for each $\varepsilon > 0$ there is a $K \in \mathcal{K}(X)$ so that $|\mu|^*(X \setminus K) < \varepsilon$. An equivalent condition is the existence of an increasing sequence (K_n) in $\mathcal{K}(X)$ so that $|\mu|^*(X \setminus K_n) \longrightarrow 0$. We denote by $M_t(X)$ the space of tight measures on X. It is clearly a vector space. If $x \in C^\infty(X)$ $\mu \in M_t(X)$, then the limit $\lim_{n\to\infty} \int x|_{K_n} d\mu_{K_n}$ exists and is independent of the particular choice of (K_n). We write $\int x\, d\mu$ for this limit.

If $K \in \mathcal{K}(X)$ and $\mu \in M(K)$, then μ defines a tight measure on X in a natural way: if $K_1 \in \mathcal{K}(X)$ and $K_1 \subseteq K$ we define μ_{K_1} to be the restriction of μ to K_1. If $K_1 \supseteq K$ we define μ_{K_1} to be the measure induced on K_1 by μ (for example, as a linear form on $C(K_1)$, μ_{K_1} is the mapping

$$x \longmapsto \int x|_{K_1} d\mu \ ).$$

Then $\tilde{\mu} := \{\mu_{K_1}\}$ is a tight measure on X and $|\tilde{\mu}|^*(X \setminus K) = 0$. Hence we can (and do) identify the space of measures on K with the subspace of $M_t(X)$ consisting of those μ for which $|\mu|^*(X \setminus K) = 0$. We denote by $M_o(X)$ the subspace $\bigcup_{K \in \mathcal{K}(X)} M(K)$ - the space of measures with <u>compact support</u>. Then $\mu \in M_o(X)$ if and only if there is a $K \in \mathcal{K}(X)$ so that $|\mu|^*(X \setminus K) = 0$.

3.2. <u>Proposition</u>: The dual of $(C^\infty(X), \tau_{\mathcal{K}})$ is naturally isomorphic to $M_o(X)$ under the bilinear form

$$(x,\mu) \longmapsto \int x\, d\mu$$

<u>Proof</u>: $(C^\infty(X), \tau_{\mathcal{K}})$ is a dense subspace of the locally convex

projective limit of the system

$$\{\rho_{K_1,K} : C(K_1) \longrightarrow C(K);\ K \subseteq K_1,\ K,K_1 \in \mathcal{K}(X)\}$$

Now by a standard result on the duals of projective limits (see SCHAEFER, Ch.I [61] § IV.4.4), the dual of the latter space is the union of the spaces $\{M(K)\}_{K \in \mathcal{K}(X)}$ i.e. $M_o(X)$ under the above identification.

3.3. Proposition: The dual of $(C^\infty(X),\beta)$ is isomorphic to the space $M_t(X)$ under the bilinear form

$$(x,\mu) \longmapsto \int x\, d\mu .$$

Proof: Each $\mu \in M_t(X)$ defines a linear form on $C^\infty(X)$ and we show that it is $\tau_{\mathcal{K}}$-continuous on the unit ball of $C^\infty(X)$. If $\varepsilon > 0$, choose $K \in \mathcal{K}(X)$ so that $|\mu|^*(X \setminus K) < \varepsilon$. Then if $x \in C^\infty(X)$ with $\|x\| \leq 1$ and $|x| \leq \varepsilon$ on K

$$\begin{aligned} \left|\int x\, d\mu\right| &\leq \left|\int_K + \int_{S\setminus K} x\, d\mu\right| \\ &\leq \varepsilon + |\mu|^*(X).\varepsilon \end{aligned}$$

and $|\mu|^*(X) < \infty$ since μ is tight.

Now let f be a β-continuous linear form on $C^\infty(X)$.

Then we can express f as a sum Σf_n where f_n is a continuous linear form on some $C(K_n)$ and $\sum_n \|f_n\| < \infty$. (I.3.10). We can regard f_n as a premeasure $\{\mu_K^n\}$ as above. Now $\{\mu_K^n\}_{n\in\mathbb{N}}$ is abolutely summable (in the Banach space $M(K)$) and so there is a $\mu_K \in M(K)$

with $\mu_K = \sum_n \mu_K^n$. It is easy to see that $\mu := \{\mu_K : K \in K(X)\}$ is a premeasure on X. If we let $K_n^1 := \bigcup_{k=1}^{n} K_k$ then

$$|\mu|^*(X \setminus K_n^1) \leq \sum_{k>n} \|f_k\|$$

and so μ is tight. One can check that $f(x) = \int x \, d\mu$ for $x \in C^\infty(X)$.

3.4. Alternative definitions of Radon measures: There are several alternative, equivalent definitions for (bounded) Radon measures on a completely regular space and, before continuing, we describe the most important of these. For convenience, we consider only non-negative measures:

A. A compact-regular Borel measure μ on X is a σ-additive finite measure on the Borel algebra of X so that for each Borel set A in X

$$\mu(A) = \sup\{\mu(K) : K \in K(X), K \subseteq A\}.$$

B. A Choquet measure on X is a bounded set function $\mu : K(X) \longrightarrow \mathbb{R}_+$ which is increasing, additive (i.e. $\mu(K_1) + \mu(K_2) = \mu(K_1 \cup K_2) + \mu(K_1 \cap K_2)$ for each pair K_1, K_2 of compacta) and continuous on the right (i.e. for $\varepsilon > 0$, $K \in K(X)$ there is a neighbourhood V of K so that $\mu(K_1) \leq \mu(K) + \varepsilon$ for each $K_1 \in K(X)$ with $K \subseteq K_1 \subseteq V$).

C. A tight measure on X is a bounded, Borel measure which satisfies the tightness condition: for every $\varepsilon > 0$, there is a compact set K so that $\mu(X \setminus K) < \varepsilon$.

D. A Radon measure μ on βX is concentrated on X if $\inf\{\mu(U) : U \text{ open and } \beta X \setminus X \subseteq U\} = 0$.

Then one can show that the above concepts all coincide in a natural way and correspond exactly to the non-negative elements of $M_t(X)$. A precise discussion can be found in SCHWARTZ [125].

In the next Proposition, we can characterise the β-equicontinuous subsets of $M_t(X)$.

3.5. Definition: We remark that if $\mu = \{\mu_K\}$ is a tight measure on X then so is the premeasure $\{|\mu_K|\}$. We denote it by $|\mu|$. A subset B of $M_t(X)$ is uniformly tight if it is bounded (for the norm) in $M_t(X)$ and satisfies the tightness condition:

for every $\varepsilon > 0$ there is a $K \in \mathcal{K}(X)$ so that $|\mu|(X \setminus K) < \varepsilon$ for each $\mu \in B$.

3.6. Proposition: A subset B of $M_t(X)$ is uniformly tight if and only if it is β-equicontinuous.

Proof: We remark firstly that it follows easily from the characterisation of equicontinuous subsets in the dual of a locally convex projective limit of Banach spaces that a subset B_1 of $M_o(X)$ is $\tau_{\mathcal{K}}$-equicontinuous if and only if it is norm bounded and has compact support (i.e. there is a $K \in \mathcal{K}(X)$ so that each $\mu \in B$ vanishes on $X \setminus K$). The result follows then from this fact and I.1.22 as in the proof of 3.3.

3.7. <u>Corollary</u>: A uniformly tight subset of $M_t(X)$ is relatively compact for the weak topology defined by $C^\infty(X)$.

Note that the converse of this result is not always true. In fact, the truth of the converse is equivalent to $(C^\infty(X),\beta)$ being a Mackey space. Hence, the first claim of 1.20 can be restated as follows:

3.8. <u>Proposition</u>: Let X be a locally compact, paracompact space. Then a weakly compact subset of $M_t(X)$ is uniformly tight.

We now consider properties of the linear operator $C^\infty(\phi)$ induced by a continuous mapping $\phi : X \longrightarrow Y$. We denote by $M_t(\phi)$ the transposed mapping of $C^\infty(\phi)$ so that $M_t(\phi)$ is a norm-bounded linear mapping from $M_t(X)$ into $M_t(Y)$. Note that if we regard a measure $\mu \in M_t(X)$ as a Borel measure (as in 3.4.A for example) then $M_t(\phi)(\mu)$ is the Borel measure

$$A \longmapsto \mu(\phi^{-1}(A))$$

i.e. it coincides with the measure induced by ϕ in the classical sense.

3.9. <u>Proposition</u>: 1) $C^\infty(\phi)$ is a quotient mapping from the Banch space $C^\infty(Y)$ onto a norm-closed subspace of $C^\infty(X)$;

2) an element μ in $M_t(Y)$ is in the range of $M_t(\phi)$ if and only if for each $\varepsilon > 0$ there is a $K \in \mathcal{K}(X)$ so that $|\mu|(Y \setminus \phi(K)) < \varepsilon$. $M_t(\phi)$ is a quotient mapping from $M_t(X)$ onto a norm-closed subspace of $M_t(Y)$;

3) $C^\infty(\phi)$ is an open mapping (for the strict topologies) from $C^\infty(X)$ onto its range if and only if $\phi(X)$ is closed in Y and for each $K \in \mathcal{K}(\phi(X))$ there is a $K_1 \in \mathcal{K}(X)$ with $\phi(K_1) \supseteq K$.

3.9 is proved by means of a series of Lemmas. To simplify the notation, we denote the operators $C^\infty(\phi)$ and $M_t(\phi)$ by U and V respectively.

3.10. Lemma: 3.9.1) holds.

Proof: We show that if $y \in C^\infty(X)$ has the form $x \circ \phi$ for some $x \in C^\infty(Y)$ then there is a $z \in C^\infty(Y)$ with $\|z\| = \|y\|$ and $y = z \circ \phi$. But this is the case for z defined as follows:

$$z(t) = \begin{cases} x(t) & \text{if } |x(t)| \leq \|y\| \\ x(t)\|y\| / |y(t)| & \text{otherwise.} \end{cases}$$

3.11. Lemma: Suppose that X and Y are compact and ϕ is surjective. Then if $\mu \in M_t(Y)$ there is a $\nu \in M_t(X)$ with $\|\mu\| = \|\nu\|$ and $V\nu = \mu$.

Proof: Since ϕ is surjective, U is an injection and so an isometry from C(Y) onto a closed subspace A of C(X). We can then regard μ as a continuous linear form on A and the result then follows by taking ν to be a Hahn-Banach extension of this functional to C(X).

3.12. Lemma: $\mu \in M_t(Y)$ is in the range of V if and only if for each $\varepsilon > 0$ there is a compact set K in X so that $|\mu|(Y \setminus \phi(K)) < \varepsilon$. Then there is a $\nu \in M_t(X)$ with $\mu = V\nu$ and $\|\nu\| = \|\mu\|$.

Proof: Necessity: suppose that $\mu = V\nu$ with $\nu \in M_t(X)$. Then for $\varepsilon > 0$, there is a $K \in \mathcal{K}(X)$, so that $|\nu|(X \setminus K) < \varepsilon$. Then clearly $|\mu|(Y \setminus \phi(K)) < \varepsilon$.

Sufficiency: we can choose an increasing sequence (K_n) of compacta in X so that $|\mu|(Y \setminus \phi(K_n)) < 1/n$. Let

$$A_1 := \phi(K_1), \quad A_n := \phi(K_n) \setminus \phi(K_{n-1}) \qquad (n > 1)$$

and put $\mu_n := \mu|A_n$ (A_n is a Borel set in Y).
Then one has $\|\mu\| = \Sigma \|\mu_n\|$.

By applying 3.11 successively to the restrictions of ϕ to K_n we get a sequence (ν_n) of Radon measures where $\nu_n \in M_t(K_n)$. $\|\nu_n\| = \|\mu_n\|$ and $V\nu_n = \mu_n$. Then the series $\Sigma \nu_n$ is absolutely summable in $M_t(X)$ and its sum ν is the required measure. In addition, we have

$$\|\nu\| \le \Sigma \|\nu_n\| = \Sigma \|\mu_n\| = \|\mu\|.$$

3.13. Lemma: If $V(M_t(X))$ is weakly closed in $M_t(Y)$, then $\phi(X)$ is closed in Y.

Proof: If $x \in C^\infty(Y)$ with $Ux = 0$, then x vanishes on $\phi(X)$ and so on $\overline{\phi(X)}$. Hence if $s \in \overline{\phi(X)}$, then δ_s is in the polar of the kernel of U. But the latter set is $V(M_t(X))$ by the bipolar theorem and so $\delta_s = V\mu$ for some $\mu \in M_t(X)$. Then

$$1 = \delta_s(\{s\}) = V\mu(\{s\}) = \mu(\phi^{-1}(s))$$

and so $\phi^{-1}(\{s\})$ is non-empty i.e. $s \in \phi(X)$.

3.14. Lemma: If every β-equicontinuous subset in $VM_t(X)$ is the image of a β-equicontinuous set in $M_t(X)$, then each $K \in \mathcal{K}(\phi(X))$ is contained in the image of a compact subset of X.

Proof: For such a K, let $B := \{\delta_s : s \in K\}$. Then B is clearly uniformly tight and so β-equicontinuous. Hence it is the image of a β-equicontinuous subset B_1 of $M_t(X)$. Then there is a K_1 in $\mathcal{K}(X)$ so that $|\mu|(X \setminus K_1) < 1/2$ for each $\mu \in B_1$. If $s \in K$ and $\mu \in B_1$ with $V\mu = \delta_s$, then

$$1 = \delta_s(\{s\}) = V\mu(\{s\}) = \mu(\phi^{-1}(\{s\}))$$

and so $\phi^{-1}(\{s\}) \nsubseteq X \setminus K_1$. Hence $\phi(K_1) \supseteq K$.

To complete the proof of 3.9, we require the following standard result on locally convex spaces:

3.15. Lemma: A continuous linear operator T from a locally convex space E into a locally convex space F is an open mapping into its range if and only if $T'(F')$ is $\sigma(E',E)$-closed in E' and each equicontinuous subset of $T'(F')$ is the image of an equicontinuous subset of F'.

Proof: See GROTHENDIECK [62].

Proof of 3.9: Only 3.9.3) remains to be proved. The necessity of the given condition follows from 3.13, 3.14 and 3.15. Sufficiency: first we note that the polar B of $VM_t(X)$ in $C^\infty(Y)$ is the set $\{x \in C^\infty(Y) : x = 0 \text{ on } \phi(X)\}$. Suppose that $\mu \in B^0$. We show that $\mu \in VM_t(X)$ and so $VM_t(X) = (VM_t(X))^{00}$ is weakly closed. There are compact sets K_n in X so that $|\mu|(X \setminus K_n) \leq 1/n$. We show that $|\mu|(K_n \setminus \phi(X)) = 0$ for each n so that

$$|\mu|(X \setminus (K_n \cap \phi(X)) \longrightarrow 0$$

which implies (by 3.12) that $\mu \in VM_t(X)$. If this were not the case, there would be an n so that

$$\delta := |\mu|(K_n \setminus \phi(X)) > 0.$$

Choose m so that $m > 2/\delta$. There is a continuous function x in $C(K_n)$ with $\|x\| = 1$, $x = 0$ on $(K_n \setminus \phi(X))$ and $\int_{K_n} x\, d\mu > \delta/2$. We can extend x without increasing the norm to a function x in $C^\infty(Y)$ which vanishes on $\phi(X)$ (and so is in the polar of $VM_t(X)$). Then

$$\int x\, d\mu > \delta/2 - 1/m > 0$$

which gives a contradiction.

A similar argument, applied to a uniformly tight set C in $VM_t(X)$ produces a uniformly tight set in $M_t(X)$ whose image is C. Hence the sufficiency follows from 3.15.

As an application of the theory developed in this section, we give a functional analytic proof of PROHOROV's theorem on the existence of projective limits of measures. We suppose that $\{\phi_{\beta\alpha} : X_\beta \longrightarrow X_\alpha, \alpha \leq \beta, \alpha,\beta \in A\}$ is a projective spectrum of completely regular spaces and that X is a completely regular space with continuous mappings $\phi_\alpha : X \longrightarrow X_\alpha$ so that $\phi_{\beta\alpha} \circ \phi_\beta = \phi_\alpha$ $(\alpha \leq \beta)$ (thus the system $\{\phi_\alpha\}$ corresponds to a continuous mapping from X into the projective limit of the system $\{X_\alpha\}$). Suppose that $\{\mu_\alpha : \alpha \in A\}$ is a compatible system of bounded Radon measures on $\{X_\alpha\}$ (i.e. $\mu_\alpha \in M_t(X_\alpha)$ and $M_t(\phi_{\beta\alpha})(\mu_\beta) = \mu_\alpha$ for $\alpha \leq \beta$). We seek necessary and sufficient conditions for the existence of a $\mu \in M_t(X)$ so that $M_t(\phi_\alpha)(\mu) = \mu_\alpha$ for each α.

3.16. <u>Proposition</u>: Such a μ exists if and only if

1) $\sup \{\|\mu_\alpha\|_{M_t(X_\alpha)} : \alpha \in A\} < \infty$;

2) for each $\varepsilon > 0$ there exists a $K \in K(X)$ so that $|\mu_\alpha|(X \setminus \phi_\alpha(K)) < \varepsilon$ for each α.

<u>Proof</u>: The necessity of condition 1) is trivial and that of 2) follows from 3.9.2).

Now suppose that 1) and 2) are satisfied. If $n \in \mathbb{N}$, choose $K_n \in K(X)$ so that $|\mu_\alpha|(X \setminus \phi_\alpha(K_n)) < 1/n$. Let

$$B := \{\mu \in M_t(X) : \|\mu\| \leq \sup \|\mu_\alpha\| \text{ and } |\mu|(X \setminus K_n) < 1/n\}$$

Then B is weakly compact in $M_t(X)$. Let

$$B_\alpha := \{\mu \in B : M_t(\phi_\alpha)(\mu) = \mu_\alpha\}$$

Then B_α is weakly compact and non-empty by 3.9.2). Hence $\cap B_\alpha \neq \phi$ by the finite intersection property.

Using 3.3 and the ideas of 2.10 - 2.11, we can give a characterisation of the dual of $(C^\infty(X),\beta_{\mathcal{B}})$ analogous to that of 3.4.C for $M_t(X)$. We denote this dual by $M_{\mathcal{B}}(X)$ (so that $M_t(X) \subseteq M_{\mathcal{B}}(X)$).

3.17. Proposition: The space $M_{\mathcal{B}}(X)$ can be naturally identified with the space of Radon measures μ on βX which satisfy the following condition: for each $\varepsilon > 0$ there is a $B \in \mathcal{B}$ so that $|\mu|(\beta X \setminus \overline{B}) < \varepsilon$ ($\overline{B}$ denotes the closure of B in βX).

Proof: By applying 3.3 to $(C^\infty(\upsilon X),\beta\ )$ we see that a $\beta_{\mathcal{B}}$-continuous linear form, which can be regarded as a $\beta_{\mathcal{K}}$-continuous linear form on $C^\infty(\upsilon X)$, is defined by a Radon measure μ on βX so that for each $\varepsilon > 0$, there is a $K \in \mathcal{K}(\upsilon X)$ with $|\mu|(\beta X \setminus K) < \varepsilon$. Since μ is $\beta_{\mathcal{B}}$-continuous, one can show as in the proof of 3.3, that one can even take K to be of the form $\overline{B}$ ($B \in \mathcal{B}(X)$).

II.4. VECTOR-VALUED FUNCTIONS

In this section, we shall assume uniformly that all spaces X are $\mathcal{S}$-complete. We recall that this means that a function x from X into $\mathbb{R}$ is continuous if its restrictions to the sets of $\mathcal{S}$ are continuous. Then the same condition holds for functions with values in a completely regular space (since such a space has the weak topology defined by the continuous, real-valued functions on it) and so, in particular, for functions with values in a locally convex space. In addition, we shall assume that $\cup \mathcal{S} = X$.

4.1. Definition: Let $(E, \| \|, \tau)$ be a complete Saks space. We denote by $C^\infty(X;E)$ the space of τ-continuous mappings x from X into E which are norm-bounded on X. On $C^\infty(X;E)$, we consider the structures:

$\| \|_E$ - the norm $x \longmapsto \sup\{\|x(t)\| : t \in X\}$

$\tau_{\mathcal{S}}^E$ - the locally convex topology of uniform τ-convergence on the sets of $\mathcal{S}$. $(C^\infty(X;E), \| \|_E, \tau_{\mathcal{S}}^E)$ is a Saks space. We denote by $\beta_{\mathcal{S}}$ the associated locally convex topology.

4.2. Proposition: 1) $\tau_{\mathcal{S}}^E \subseteq \beta_{\mathcal{S}} \subseteq \tau_{\| \|_E}$

2) $(C^\infty(X;E), \beta_{\mathcal{S}})$ is complete;

3) a subset $B \subseteq C^\infty(X;E)$ is $\beta_{\mathcal{S}}$-bounded if and only if it is norm-bounded;

4) a sequence (x_n) in $C^\infty(X;E)$ converges to zero for $\beta_{\mathcal{S}}$ if and only if it is norm-bounded and $\tau_{\mathcal{S}}^E$-convergent to zero;

5) a linear operator from $C^\infty(X;E)$ into a locally convex space is $\beta_{\mathcal{S}}$-continuous if and only if its restriction to the unit ball of $(C^\infty(X;E), \| \|_E)$ is $\tau_{\mathcal{S}}^E$-continuous;

6) if E is a Banach space and X is $\mathcal{S}$-$\mathcal{S}$ normal, then $\beta_{\mathcal{S}}$ is defined by the seminorms $\{p_\phi : \phi \in L_{\mathcal{S}}^+\}$ where

$$p_\phi : x \longmapsto \|\phi x\|_E$$

$C^{\infty}(X;E)$, as a Saks space, has a natural projective limit representation which we now display. Suppose that E is the Saks space projective limit of the sequence

$$\{\pi_{qp} : \hat{E}_q \longmapsto \hat{E}_p, \ p \leq q, \ p,q \in P\}$$

where P is a suitable family of seminorms which defines the topology τ. Then we can order the set $S \times P$ in a natural way and if $(B,p) \leq (B_1,q)$ we denote by $\rho^{q,p}_{B_1,B}$ the mapping

$$x \longmapsto \pi_{qp}(x|_B)$$

from the Banach space $C^{\infty}(B_1;\hat{E}_q)$ into $C^{\infty}(B;\hat{E}_p)$. Then

$$\{\rho^{q,p}_{B_1,B} : C^{\infty}(B_1;\hat{E}_q) \longrightarrow C^{\infty}(B;\hat{E}_p); \ p \leq q, \ B \subseteq B_1\}$$

is a projective system of Banach space and $C^{\infty}(X;E)$ can be identified with its Saks space projective limit in a natural way.

Now if X is compact and E is a Banach space, there is a natural isometry from $C(X) \hat{\hat{\otimes}} E$ onto $C(X;E)$ (see, for example, SEMADENI [127], § 20.5.6) which is constructed as follows:
if $x = \Sigma x_i \otimes a_i$ ($x_i \in C(X)$, $a_i \in E$) we define Jx to be the function

$$t \longmapsto \Sigma x_i(t) a_i$$

from X into $\mathbb{C}$. Then J is an isometry from $C(X) \otimes E$ (with the inductive tensor product norm) onto a dense subspace of $C(X;E)$ and so extends to the desired isometry. We can now easily obtain the following generalisation of this result.

4.3. Proposition: The Saks space $C^\infty(X;E)$ is naturally isomorphic to the tensor product $(C^\infty(X) \hat{\hat{\otimes}}_\gamma E, \| \|, \tau_S \hat{\hat{\otimes}} \tau)$ provided that $S \subseteq K$.

Proof: Consider the commutative diagram

$$\begin{array}{ccccc} C^\infty(X;E) & \longrightarrow & C(B_1;\hat{E}_q) & \longrightarrow & C(B;\hat{E}_p) \\ \downarrow & & \downarrow & & \downarrow \\ C^\infty(X) \hat{\hat{\otimes}}_\gamma E & \longrightarrow & C(B_1) \hat{\hat{\otimes}} \hat{E}_q & \longrightarrow & C(B) \hat{\hat{\otimes}} \hat{E}_p \end{array}$$

where the vertical arrows are the natural isometries desribed above (we are using I.3.12 to identify $C^\infty(X) \hat{\hat{\otimes}}_\gamma E$ with the Saks space projective limit of the system defined by the spaces $\{C(B) \hat{\hat{\otimes}} E_p\}$. The required isomorphism is the unique vertical arrow from $C^\infty(X;E)$ into $C^\infty(X) \hat{\hat{\otimes}}_\gamma E$ which preserves the commutativity.

Now let X and X_1 be completely regular spaces with suitable saturated families S and S_1 where $S \subseteq K(X)$ and $S_1 \subseteq K(X_1)$. We define a saturated family $S \times S_1$ on the product space $X \times X_1$, consisting of those closed subsets of $X \times X_1$ which are contained in a set of the form $B \times B_1$ ($B \in S$, $B_1 \in S_1$). Note that $X \times X_1$ need not be $S \times S_1$-complete, in general. We denote its $S \times S_1$-completion (as described in 1.7) by Y.

If x is a mapping from X into $C^\infty(X_1)$ then we define Jx to be the mapping

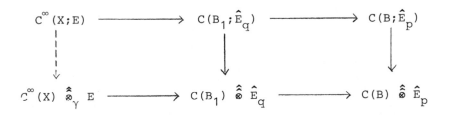

$$(s,s_1) \longmapsto \{x(s)\}(s_1)$$

on $X \times X_1$. If we regard $C^\infty(X)$ and $C^\infty(X_1)$ as Banach spaces (with the supremum norm), it is well-known that $J : x \longmapsto Jx$ is an isometry (cf. SEMADENI [127], § 5.3.4) - in general, it is not surjective but this <u>is</u> the case if X and X_1 are compact (loc. cit. § 7.7.5 - in fact, it suffices to require that X_1 be compact). The following result is a natural generalisation of this fact to non-compact spaces:

4.4. <u>Proposition</u>: J induced a Saks space isomorphism from $C^\infty(X;C^\infty(X_1))$ onto $C^\infty(Y)$.

<u>Proof</u>: Firstly, it follows from the result for compact spaces that if $x \in C^\infty(X;C^\infty(X_1))$ then Jx is continuous on sets of the form $B \times B_1$ ($B \in \mathcal{S}$, $B_1 \in \mathcal{S}_1$) and so Jx lies in $C^\infty(Y)$. The image of $C^\infty(X;C^\infty(X_1))$ is a selfadjoint subalgebra of $C^\infty(Y)$, containing the constants, and so is $\beta_{\mathcal{S}\times\mathcal{S}_1}$-dense in $C^\infty(Y)$ by the complex version of 1.14. Since $C^\infty(X;C^\infty(X_1))$ is a complete Saks space it is sufficient to verify that its image is a Saks-subspace of $C^\infty(Y)$ and this follows from the results mentioned above.

4.5. <u>Corollary</u>: If X and X_1 are locally compact then J induces an isomorphism from $(C^\infty(X;C^\infty(X_1)),\beta_K)$ onto $(C^\infty(X \times X_1),\beta_K)$.

Combining 4.3 and 4.4 we get

4.6. Proposition: There is a natural Saks space isomorphism from $(C^\infty(Y), \|\ \|, \tau_{S\times S_1})$ onto $(C^\infty(X) \hat{\hat{\otimes}}_\gamma C^\infty(X_1), \|\ \|, \tau_S \hat{\hat{\otimes}} \tau_{S_1})$.

In § 2 we studied $C^\infty(X)$ as an algebra. $C^\infty(X;E)$ does not, in general have the structure of an algebra but it is, in a natural way, a $C^\infty(X)$-module. In the next Proposition, we characterise the closed submodules of $C^\infty(X;E)$. Let M be such a submodule i.e. it is a vector subspace with the property that $xM \subseteq M$ for each $x \in C^\infty(X)$. We then define, for $s \in X$,

$$M_s := \overline{\{x(s) : x \in M\}} \qquad \text{(closure in } \beta)$$

i.e. M_s is the β-closure of $M(s)$ in E. Then

$$\tilde{M} := \{x \in C^\infty(X;E) : x(s) \in M_s \quad \text{for each } s \in X\}$$

is obviously a β-closed submodule of $C^\infty(X;E)$. We show that $M = \tilde{M}$ if (and only if) M is β-closed.

4.7. Lemma: Suppose that X is compact and E is a Banach space. Then if M is a (norm)-closed submodule of $C(X;E)$, $M = \tilde{M}$.

Proof: The result follows from the following formula which holds for any $x \in C^\infty(X;E)$:

$$\inf_{y\in M} \|x - y\| = \sup_{s\in X} \inf_{z\in M_s} \|x(s) - z\|$$

The left-hand side is clearly greater than the right-hand side. Now choose $\varepsilon > 0$. Then for $s \in X$ there is a $u \in M$ so that

$$\|x(s) - u(s)\| < \sup_{s\in X} \inf_{z\in M_s} \|x(s) - z\| + \varepsilon$$

This inequality holds on an open neighbourhood of s in X. Hence, by compactness, we can cover X by open sets $\{U_1,\ldots,U_n\}$ to which there correspond functions $u_1,\ldots,u_n$ in M so that

$$\|x(s) - u_i(s)\| \le \sup_{s\varepsilon X} \inf_{z\varepsilon M_s} \|x(s) - z\| + \varepsilon$$

on U_i $(i = 1,\ldots,n)$. Let $\{\phi_1,\ldots,\phi_n\}$ be a partition of unity subordinate to $\{U_i\}$ and put $y := \Sigma\, \phi_i u_i$. Then a routine calculation shows that

$$\|x(s) - y(s)\| \le \sup_{s\varepsilon X} \inf_{z\varepsilon M_s} \|x(s) - z\| + \varepsilon \quad (s \ \varepsilon\ X)$$

so that the above equality holds as claimed.

4.8. Proposition: If X is a completely regular space, E a complete Saks space and M a submodule of $C^{\infty}(X;E)$, then $M = \widetilde{M}$ if and only if M is β-closed.

Proof: Suppose that M is β-closed. We show that $M = \widetilde{M}$. For $y \ \varepsilon\ \widetilde{M}$, we construct a net $(y_{K,\alpha})$ indexed by $K(X) \times A$ (where A is the indexing set in a projective limit representation $\{\pi_{\beta\alpha} : E_\beta \longmapsto E_\alpha\}$ of E) which is β-convergent to y.

Let $M_{K,\alpha}$ be the projection of M in $C(K;E_\alpha)$. Then $M_{K,\alpha}$ is a $C(K)$-submodule of $C(K;E_\alpha)$. Now it is clear that if $s \ \varepsilon\ K$, then $M_{K,\alpha}(s) = \pi_\alpha(M(s))$ where π_α is the natural mapping from E into E_α. Hence $\pi_{K,\alpha}(y) \ \varepsilon\ \widetilde{M}_{K,\alpha}$ and so is approximable (uniformly on K with respect to the norm of E_α) by a function in M (by 4.7). By a standard argument, we can ensure that these approximations are uniformly bounded.

We remark that 4.8 contains Proposition 2.7 as a special case.

II.5. GENERALISED STRICT TOPOLOGIES

In this section, we consider a completely regular space X as a subspace of its Stone-Čech compactification βX. $\mathcal{K}(\beta X \setminus X)$ denotes the collection of compact subsets of $\beta X \setminus X$. If $K \in \mathcal{K}(\beta X \setminus X)$ then $\beta X \setminus K$ is a locally compact space containing X. The vector spaces $C^\infty(X)$ and $C^\infty(\beta X \setminus K)$ are naturally isomorphic (under the restriction mapping from $C^\infty(\beta X \setminus K)$ onto $C^\infty(X)$). We denote by β_K the strict topology on $C^\infty(\beta X \setminus K)$ and we regard it as a locally convex topology on $C^\infty(X)$. For reasons which will become clear later, we denote the topology $\beta_{\mathcal{K}}$ on $C^\infty(X)$ by β_t in this section. Then β_t is coarser than each β_K.

Now suppose that $\mathcal{L}$ is a subfamily of $\mathcal{K}(\beta X \setminus X)$. For convenience, we shall assume that $\mathcal{L}$ is closed under the formation of finite unions. If $K \subseteq K_1$ then β_K is finer than β_{K_1}. Hence the family $\{\beta_K : K \in \mathcal{L}\}$ forms a directed system of locally convex topologies on $C^\infty(X)$. We can then define the locally convex topology $\beta_{\mathcal{L}}$ on $C^\infty(X)$ to be the locally convex inductive limit of these topologies i.e. the finest locally convex topology on $C^\infty(X)$ which is coarser than each β_K $(K \in \mathcal{L})$. $\beta_{\mathcal{L}}$ is then finer than β_t. The following choices of $\mathcal{L}$ will be especially important:

a) $\mathcal{L} = \{\phi\}$ - then $\beta_{\mathcal{L}}$ is the uniform topology;

b) $\mathcal{L}_\sigma$ - the family of sets in $\mathcal{K}(\beta X \setminus X)$ which are zero-sets of functions in $C(\beta X)$. Note that $K \in \mathcal{K}(\beta X \setminus X)$ is in $\mathcal{L}_\sigma$ if and only if $\beta X \setminus K$ is σ-compact. For if $K = Z(x)$ $(x \in C(\beta X))$

then

$$\beta X \setminus K = \bigcup_{n \in \mathbb{N}} \{s : |x(s)| \geq 1/n\}$$

and each of these sets is compact. On the other hand, if $\beta X \setminus K$ is expressible as a union $\bigcup_{n \in \mathbb{N}} K_n$ of compact sets where $K_n \subseteq K_{n+1}^o$ for each n, then a standard construction produces a continuous function x on $\beta X \setminus K$ so that $|x| \leq 1/n$ on $(\beta X \setminus K) \setminus K_n$ and the extension of x to βX has K as zero set. The topology on $C^\infty(X)$ defined by L_σ will be denoted by β_σ;

c) L_τ - the family of all compact subsets of $\beta X \setminus X$. The corresponding topology on $C^\infty(X)$ will be denoted by β_τ.

We note first that each of the topologies β_L is a mixed topology. Of course, the unit ball of $C^\infty(X)$ is closed in each β_L (since β_L is finer than the topology of pointwise convergence on X).

5.1. Proposition: $\beta_L = \gamma[\| \|, \beta_L]$.

Proof: It suffices to show that any linear operator from $C^\infty(X)$ into a locally convex space F is β_L-continuous when it is β_L-continuous on $B_{\| \|_\infty}$. But then it is β_K-continuous on $B_{\| \|_\infty}$ and so β_K-continuous. Continuity then follows from the universal property of inductive limits.

5.2. Proposition: a) $\beta_t \subseteq \beta_\tau \subseteq \beta_\sigma$;
b) $\beta_t = \beta_\tau$ if X is locally compact.

Proof: a) is trivial. b) follows from the fact that X is locally compact if and only if it is open in βX i.e. if and only if $\beta X \setminus X \in \mathcal{L}_\tau$.

One of the most interesting and important problems regarding strict topologies is that of characterising in purely topological terms those spaces for which given strict topologies coincide. One simple method of obtaining necessary conditions is to characterise the spectrum of $(C^\infty(X), \beta_{\mathcal{L}})$.

5.3. Proposition: The spectrum of $(C^\infty(X), \beta_{\mathcal{L}})$ can be identified via the generalised Dirac transform with $\bigcap_{K \in \mathcal{L}} \beta X \setminus K$.

Proof: An element of the spectrum has the form δ_s for some $s \in \beta X$. Now δ_s is in the $\beta_{\mathcal{L}}$-spectrum if and only if it is β_K-continuous for each $K \in \mathcal{L}$ i.e. if and only if $s \in \beta X \setminus K$ for each $K \in \mathcal{L}$ (2.2).

5.4. Corollary: a) The spectrum of $(C^\infty(X), \beta_\sigma)$ is υX, the realcompactification of X;

b) the spectrum of $(C^\infty(X), \beta_\tau)$ is X;

c) if $\beta_\sigma = \beta_\tau$, then X is realcompact.

Proof: a) follows immediately from 5.3 and the definition of υX. b) is trivial and c) follows from a) and b).

5.5. Proposition: The space $(C^\infty(X),\beta_\sigma)$ is a Mackey space for any X.

Proof: By 1.20, $(C^\infty(X),\beta_K)$ is Mackey for each $K \in L_\sigma$ and the inductive limit of Mackey spaces is Mackey.

5.6. Proposition: For any X, the following are equivalent:

a) X is compact;

b) $(C^\infty(X),\beta_\tau)$ is barrelled (bornological, metrisable or normable).

Proof: By I.1.15, any of the conditions of b) is equivalent to the condition that $\beta_\tau = \tau_{\|\ \|_\infty}$ i.e. that $\beta_K = \tau_{\|\ \|_\infty}$ for each $K \in L_\tau$. Clearly, this can only happen if $L_\tau = \{\phi\}$ i.e. $X = \beta X$.

Recall that a completely regular space X is pseudocompact if $C(X) = C^\infty(X)$ (i.e. if X is bounded in itself).

5.7. Proposition: For any X, the following conditions are equivalent:

a) X is pseudocompact;

b) β_σ is barrelled (bornological, metrisable, normable).

Proof: Using the simple remark that X is pseudocompact if and only if $\upsilon X = \beta X$, one can prove this as in 5.6.

We now define, using a rather different method, a fourth strict topology β_∞ on $C^\infty(X)$. We denote by $\mathcal{D}$ the family of all continuous pseudometrics on X. If $d \in \mathcal{D}$ then we denote by X_d the associated metric space (i.e. the quotient of X with respect to the equivalence relation $x \sim y$ if and only if $d(x,y) = 0$). Then there are natural projections

$$\pi_d : X \longrightarrow X_d$$

and, dualising these, we obtain injections

$$C^\infty(\pi_d) : C^\infty(X_d) \longrightarrow C^\infty(X)$$

(i.e. $C^\infty(\pi_d)$ identifies $C^\infty(X_d)$ with the subspace of those functions in $C^\infty(X)$ which factorise over π_d).

Since X_d is a completely regular space (with the topology induced by the metric), we can supply $C^\infty(X_d)$ with the strict topology β_τ and so are in a position to topologise $C^\infty(X)$ by defining β_∞ to be the locally convex inductive limit of these topologies.

5.8. <u>Proposition</u>: a) $\beta_\tau \subseteq \beta_\infty$;

b) β_∞ is a mixed topology i.e. $\beta_\infty = \gamma[\|\ \|, \beta_\infty]$.

<u>Proof</u>: It is easy to show that $C^\infty(\pi_d)$ is β_τ-continuous from $C^\infty(X_d)$ into $C^\infty(X)$ (since $\pi_d : X \longrightarrow X_d$ is continuous). This proves a).

b) cf. the proof of 5.1.

5.9. Remark: I. WHEELER [152] has shown that $\beta_\infty \subseteq \beta_\sigma$.

II. The topology β_∞ is, in fact, a β_L, although for a rather awkard choice of L.
Let $(x_\alpha)_{\alpha\varepsilon A}$ be a net in $C^\infty(X)$ which is uniformly bounded, equicontinuous and decreases to zero. If $\varepsilon > 0$ put

$$K := \{s \varepsilon \beta X : x_\alpha(s) \geq \varepsilon \quad \text{for all } \alpha \varepsilon A\}$$

Then if we define L to be the set of all K which can be described in this manner, $\beta_\infty = \beta_L$.

III. The above construction is rather artificial in the context of completely regular spaces and depends more on the fine uniform structure of X (i.e. the finest uniform structure compatible with the topology or, alternatively, the projective uniform structure induced by the $\{\pi_d\}$) than on its topology. In fact, we shall see in the Appendix that the space $(C^\infty(X),\beta_\infty)$ arises more naturally in the context of a duality theory for uniform spaces.

In the light of this remark, it is not surprising that the spectrum of the algebra $(C^\infty(X),\beta_\infty)$ is the topological completion of X (i.e. the completion of X with respect to the fine uniformity). This will also be proved in the Appendix as a special case of a result on uniform spaces.

The significance of the topologies β_σ and β_τ on $C^\infty(X)$ is that their dual spaces are identifiable with important spaces of measures on X - the σ-additive Baire measures and the τ-additive

Borel measures respectively. We recall briefly the details. If X is a completely regular space, we denote by

- Ba(X) - the family of Baire sets in X i.e. the σ-algebra generated by the zero sets in X;
- Bo(X) - the family of Borel sets i.e. the σ-algebra generated by the closed sets.

$M_f(X)$ denotes the space of finite, finitely additive, complex-valued Baire measures on X;
$M_\sigma(X)$ denotes the space of finite, σ-additive, complex-valued Baire measures on X;
$M_\tau(X)$ denotes the space of finite, τ-additive, complex-valued Borel measures on X.

The basis for the identification between measures as linear forms, resp. as set functions, is the following classical result:

5.10. Proposition: a) Integration establishes an order-preserving vector isomorphism from $M_f(X)$ onto the dual space of $(C^\infty(X), \| \|)$;

b) under this bijection, $M_\sigma(X)$ corresponds to the set of σ-additive functionals on $C^\infty(X)$ i.e. those f which satisfy the condition: $f(x_n) \longrightarrow 0$ for each sequence (x_n) in $C^\infty(X)$ which is uniformly bounded and decreases to zero;

c) under this bijection, $M_\tau(X)$ corresponds to the set of τ-additive functionals on $C^\infty(X)$ i.e. those f which satisfy the condition: $f(x_\alpha) \longrightarrow 0$ for each net $(x_\alpha)_{\alpha\varepsilon A}$ in $C^\infty(X)$ which is uniformly bounded and decreases to zero.

For details, we refer to VARADARAJAN [144].

Note that if $\mu \varepsilon M_f(X)$, then μ induces a norm continuous linear functional on $C(\beta X)$ (= $C^\infty(X)$) and so can be regarded as a

Baire measure on βX which we denote by $\bar{\mu}$;

Borel measure on βX which we denote by $\bar{\nu}$;

We denote the inner measures associated to $|\bar{\mu}|$ and $|\bar{\nu}|$ by $|\bar{\mu}|_*$, $|\bar{\nu}|_*$ resp. Hence, for $A \subseteq \beta X$,

$$|\bar{\mu}|_*(A) = \sup\{ |\bar{\mu}|(K) : K \subseteq A, K \text{ a zero set}\}$$

$$|\bar{\nu}|_*(A) = \sup\{ |\bar{\nu}|(K) : K \subseteq A, K \varepsilon \mathcal{K}(\beta X)\}.$$

The following result characterises σ- and τ-additivity in terms of the measure of the remainder space $\beta X \setminus X$ with respect to these measures.

5.11. Proposition: For a measure $\mu \varepsilon M_f(X)$, the following are equivalent:

a) $\mu \varepsilon M_\sigma(X)$;

b) $|\bar{\mu}|_*(\beta X \setminus X) = 0$;

c) $|\bar{\mu}|(K) = 0$ for each zero set in $\beta X \setminus X$.

Proof: The equivalence of b) and c) follows immediately from the definition of $|\bar{\mu}|_*$.

a) $\Longrightarrow$ c): let K be a zero-set in $\beta X \setminus X$. One can construct a sequence (x_n) in $C(\beta X)$ which is uniformly bounded and decreases to χ_K, the characteristic function of K. Then (x_n) decreases to zero on X and so, by σ-additivity, $|\bar{\mu}|_*(K) = 0$.

c) $\Longrightarrow$ a): let (x_n) be a uniformly bounded sequence in $C^\infty(X)$ which decreases to zero on X. It is no loss of generality to suppose that $0 \leq x \leq 1$. Choose $\varepsilon > 0$ and put

$$K_n := \{s \in \beta X : x_n(s) \geq \varepsilon\}$$

Then (K_n) is a sequence of zero sets in βX and $\bigcap K_n$ is a zero set in $\beta X \setminus X$ and so has zero $|\bar{\mu}|$-measure. It follows then from σ-additivity that $|\bar{\mu}|(K_n) \longrightarrow 0$. But

$$\int_X x_n \, d|\mu| = \int_{\beta X} x_n \, d|\bar{\mu}| \leq |\bar{\mu}|(K_n) + \varepsilon|\bar{\mu}|(\beta X)$$

and so $\int_X x_n \, d|\mu| \longrightarrow 0$ i.e. $|\mu| \in M_\sigma(X)$ and so $\mu \in M_\sigma(X)$.

The next result is proved similary:

5.12. <u>Proposition</u>: For a measure $\mu \in M_f(X)$, the following are equivalent:

a) $\mu \in M_\tau(X)$;

b) $|\bar{\nu}|_*(\beta X \setminus X) = 0$;

c) $|\bar{\nu}|(K) = 0$ for each $K \in \mathcal{K}(\beta X \setminus X)$.

Using this result, we can now identify the duals of $C^\infty(X)$ under β_σ and β_τ. Firstly we characterise the dual of $(C^\infty(X), \beta_K)$

($K \varepsilon \mathcal{K}(\beta X \setminus X)$). Now $\beta X \setminus K$ is an open (and hence locally compact) subset of βX and it is classical that the Radon measures on $\beta X \setminus K$ coincide with the Radon measures on βX which have support in $\beta X \setminus K$ (see, for example, BOURBAKI [14]).

5.13. <u>Lemma</u>: Let $K \varepsilon \mathcal{K}(\beta X \setminus X)$. Then the dual of $(C^{\infty}(X), \beta_K)$ is naturally identifiable with the set of those $\mu \varepsilon M_f(X)$ which satisfy the condition $|\bar{\mu}|(K) = 0$.

5.14. <u>Proposition</u>: The dual of $(C^{\infty}(X), \beta_{\mathcal{L}})$ is naturally identifiable with the space of those $\mu \varepsilon M_f(X)$ for which $|\bar{\mu}|(K) = 0$ for each $K \varepsilon \mathcal{L}$.

Combining 5.11, 5.12, and 5.14, we get:

5.15. <u>Proposition</u>: a) The dual of $(C^{\infty}(X), \beta_{\sigma})$ is naturally identifiable with $M_{\sigma}(X)$;

b) the dual of $(C^{\infty}(X), \beta_{\tau})$ is naturally identifiable with $M_{\tau}(X)$.

We now consider some results which establish a relationship between the topological properties of a space and measure-theoretical properties. As a simple example of such a result note that X is pseudocompact if and only if $M_f(X) = M_{\sigma}(X)$. For since β_{σ} is the Mackey topology (5.5), the latter equality holds exactly when $\beta_{\sigma} = \tau_{\| \|_{\infty}}$ and this means that X is pseudocompact (5.7).

In the following, we shall concentrate on three measure theoretical properties of X - the Prohorov condition, measure compactness and strong measure compactness.

5.16. Definition: A completely regular space X is said to be a *T-space* if every weakly compact subset of $M_t^+(X)$ is β_t-equicontinuous (the superscript + denotes the set of non-negative measures) (alternative names - a *Prohorov space* or *space which satisfies Prohorov's condition*). The following result is classical (see BOURBAKI [14], § IX.5.5, Th. 2):

5.17. Proposition: If X is locally compact, then X is a T-space.

5.18. Corollary: For any suitable family L, a weakly compact subset of non-negative measures in $(C^\infty(X),\beta_L)'$ is β_L-equicontinuous. In particular, a weakly compact subset of $M_\sigma^+(X)$ (resp. $M_\tau^+(X)$) is β_σ-equicontinuous (resp. β_τ-equicontinuous).

5.19. Proposition: Let X,Y be completely regular spaces and suppose that there exists a continuous mapping $\phi : X \longrightarrow Y$ with the property that if $K \in K(Y)$, then $\phi^{-1}(K) \in K(X)$. Then if Y is a T-space, so is X.

Proof: Suppose that $B \subseteq M_t^+(X)$ is weakly compact. Then so is $C^\infty(\phi)(B)$ $(\subseteq M_t^+(Y))$ and so, for $\varepsilon > 0$, there is a $K \in K(Y)$ with

$$C^\infty(\phi)(\mu)(Y \setminus K) < \varepsilon$$

for each $\mu \in B$. But then $\mu(X \setminus \phi^{-1}(K)) < \varepsilon$ for each μ and so B is equicontinuous.

5.20. <u>Corollary</u>: A closed subspace of a T-space is a T-space.

5.21. <u>Proposition</u>: Let (X_n) be a sequence of T-spaces. Then $X := \prod_{n \in \mathbb{N}} X_n$ is a T-space.

<u>Proof</u>: Let B be a weakly compact subset of $M_t^+(X)$. We can suppose that B lies in the unit ball of $M_t(X)$. If $\varepsilon > 0$, then for each $n \in \mathbb{N}$, there is a $K_n \in \mathcal{K}(X_n)$ so that

$$C^\infty(\pi_n)(\mu)(X_n \setminus K_n) < \varepsilon/2^n \qquad (\mu \in B)$$

(π_n denotes the natural projection from X onto X_n). Then if $K := \prod_{n \in \mathbb{N}} K_n$, a simple calculation shows that $\mu(X \setminus K) < \varepsilon$ for each $\mu \in B$.

5.22. <u>Corollary</u>: a) Let (X_n) be a sequence of T-subspaces of a space X. Then $\bigcap X_n$ is a T-space;

b) a G_δ-subspace of a compact space is a T-space;

c) a complete metrisable space is a T-space.

<u>Proof</u>: a) follows from the simple fact that $\bigcap X_n$ is homeomorphic to a closed subset of $\prod X_n$ (via the diagonal mapping). We can then apply 5.20 and 5.21.

b) follows from a) and the fact that an open subset of a compact

space, being locally compact, is a T-space (5.17).
c) follows from b) and the fact that a metrisable space possesses a compatible, complete metric if and only if it is a G_δ-subset of its Stone-Čech compactification (see, for example, COMFORT and NEGREPONTIS [35], Theorem 3.6).

5.23. Remark: We mention without proof some further results on T-spaces. An example of a non-T-space (which is even Lusinian i.e. possesses a finer separable, complete metrisable topology) is an infinite dimensional separable Hilbert space, provided with the weak topology (this example is due to FERNIQUE - see BADRIKIAN [2], Exp. 8). The proof of 5.22.a) shows that every X which is a G_δ in βX is a T-space. Such spaces are said to be topologically complete (or complete in the sense of Čech). PREISS [115] has shown that a separable, co-analytic metric space is a T-space if and only if it is topologically complete. In particular, $\mathbb{Q}$ is not a T-space (we remark that in [114], VARADARAJAN had claimed to prove that a metric space is always a T-space). MOSIMAN and WHEELER [105] have shown that a space X is a T-space provided

a) it is the locally-finite union of closed T-subspaces

or b) it has an open cover consisting of T-spaces.

(see also HOFFMANN-JØRGENSEN [80]).

5.24. Definition: A completely regular space X is said to be measure compact if $M_\sigma(X) = M_\tau(X)$.

5.25. Proposition: a) If X is measure-compact, then it is real-compact;

b) X is measure-compact if and only if $\beta_\sigma = \beta_\tau$.

Proof: a) follows immediately from 5.4 and b) from the fact that β_σ is the Mackey topology.

5.26. Remark: The converse of 5.25.a) does not hold (despite several claims to the contrary which have been published). In fact, MORAN [101] has shown that the Sorgenfrey plane is a real-compact space which is not measure-compact. He has also shown that paracompactness is not sufficient for measure-compactness.

For the next result, we recall that a space X is Lindelöf if every open cover of X has a countable subcover. This property can be characterised in terms of the embedding of X in βX (cf. COMFORT and NEGREPONTIS [35], Th. 2.3).

5.27. Proposition: X is Lindelöf if and only if for every compact $K \subseteq \beta X \setminus X$ there is a zero-set Z of βX so that $K \subseteq Z \subseteq \beta X \setminus X$.

Proof: Suppose that X is Lindelöf and K as above. For each $s \in X$ there is a continuous $x_s : \beta X \longrightarrow [0,1]$ which vanishes on K and takes on the value 1 at s. We can find a sequence (s_n) in X so that $X \subseteq \bigcup_{n \in \mathbb{N}} x_{s_n}^{-1}(]1/2, 1])$. Then $x := \sum_n 2^{-n} x_{s_n}$ is continuous on βX and its zero-set satisfies the required condition.

On the other hand, if $\mathcal{U}$ is an open cover of X and if, for each $U \in \mathcal{U}$, U' is an open set in βX with $U = U' \cap \beta X$, then $\bigcup \{U' : U \in \mathcal{U}\}$ contains a set of the form $\operatorname{coz} x$ ($\supseteq X$) for some $x \in C(\beta X)$. Now $\operatorname{coz} x$ is σ-compact and so is covered by countably many of the $\{U'\}$. X is then covered by the corresponding U's.

5.28. Proposition: If X is Lindelöf, then $\beta_\sigma = \beta_\tau$ and so X is measure compact.

Proof: This follows immediately from the fact that if X is Lindelöf then $\{C^\infty(\beta X \setminus K) : K \in \mathcal{L}_\sigma\}$ is cofinal in $\{C^\infty(\beta X \setminus K) : K \in \mathcal{L}_\tau\}$.

5.29. Example: We sketch briefly an example of a non-measure compact space. Let X be the space of ordinals less than the first uncountable ordinal, provided with the order topology. It is known that $x \in C^\infty(X)$ is eventually constant i.e. there is an $s_0 \in X$ so that $x(s) = x(s_0)$ for $s \geq s_0$. Then the functional

$$x \longmapsto x(s_0)$$

is obviously not in $M_\tau(X)$. However, it is σ-additive since no countable subset of X is co-final (see GILLMAN and JERISON [55], §§ 5.11-5.13 for the properties of X that we have used here).

5.30. Remark: We mention without proof the following stability properties of measure compact spaces. A closed or Baire subspace of a measure compact space is measure compact. In general, finite products of measure compact spaces need not be measure compact but this does hold (even for countable products) when local-compactness is assumed (for details, see MORAN [102], MOSIMAN and WHEELER [105] and KIRK [88]).

The next condition that we consider is the equality $M_\tau(X) = M_t(X)$. We can characterise this property as follows:

5.31. Proposition: $M_\tau(X) = M_t(X)$ if and only if X is absolutely Borel measurable in βX.

Proof: We recall that the second condition means that for any compact regular Borel measure μ on βX, X is the union of a Borel set in βX and a μ-negligible set.
The proof follows from 5.12.

5.32. Corollary: If X is locally compact or complete metrisable, then $M_\tau(X) = M_t(X)$.

Proof: For then X is open (resp. a G_δ-set) in βX and so is absolutely Borel measurable.

5.33. Definition: A space is strongly measure compact if $M_\sigma(X) = M_t(X)$.

5.34. <u>Proposition</u>: A σ-compact space or a Polish space (i.e. separable, complete metrisable space) is strongly measure compact.

<u>Proof</u>: Such spaces are clearly Lindelöf and absolutely Borel measurable in βX so that we can apply 5.28 and 5.31.

5.35. <u>Remark</u>: KNOWLES [92], p. 149 has shown that the standard non-measurable subset of [0,1] is not strongly measure compact. Hence this space, being separable metrisable, is an example of a measure compact space which is not strongly measure compact. MORAN [102], Def. 4.1 and Prop. 4.4, has given the following characterisation of strongly measure compact spaces: X is strongly measure compact if and only if for each $\mu \neq 0$ in $M_\sigma^+(X)$, there is a $K \in \mathcal{K}(X)$ with $\mu(K) > 0$.

Using this result, one can show that the class of strongly measure compact spaces is closed under the following operations:

a) countable products;

b) Baire or closed subsets;

c) countable intersection.

5.36. <u>Definition</u>: A completely regular space X is β-<u>simple</u> if $\beta_\sigma = \beta_t$. The next result follows directly from the Definition, 5.5 and 5.18.

5.37. Proposition: Let X be a regular space. Then the following statements are equivalent:

a) X is β-simple;

b) X is a strongly measure compact T-space;

c) X is strongly measure compact and β_t is the Mackey topology;

d) every weakly compact subset of $M_\sigma(X)$ is β_t-equicontinuous;

e) every weakly compact subset of $M_\sigma^+(X)$ is β_t-equicontinuous.

5.38. Corollary: If X is a Polish space or a σ-compact, locally compact space, then X is β-simple.

5.39. Remark: The class of β-simple spaces is closed under the following operations (see MOSIMAN and WHEELER [105], Th. 6.1):

a) closed subspaces;

b) countable products and intersection.

II.6. NOTES

II.1. The study of strict topologies on $C^\infty(X)$ was initiated almost simultaneously by BUCK [26] and [27], LE CAM [94] and MAŘIK [98]. Mařik and Buck used weighted semi-norms and Le Cam used a mixed topology. The fact that the two approaches were identical was noted later by several authors (see e.g. COOPER [39], DORROH [45], FREMLIN et al. [50] and WEBB [147]). Almost all of the results in this section are natural generalisations and strengthenings of results of BUCK [27]. For studies of strict topologies see COLLINS and DORROH [32], FREMLIN et al. [50], GILES [54], HOFMANN-JØRGENSEN [79], VAN ROOIJ [120] and SENTILLES [132]. For Stone-Weierstraß theorems for the strict topology, see BUCK [27], FREMLIN et al. [50], GILES [54], GLICKSBERG [56], HAYDON [75], HOFMANN-JØRGENSEN [79], TODD [140] and WELLS [150]. This can be regarded as a special case of the weighted approximation problem (see BIERSTEDT [6], MACHADO and PROLLA [97], NACHBIN [106] and NACHBIN, MACHADO and PROLLA [107]). Separability of the strict topology has been studied by GULICK and SCHMETS [68], SCHMETS [121] and SUMMERS [139]. In the proof of 1.15 we have used an idea of WARNER [146] (who considered the essentially identical problem of separability under the topology of compact convergence).

The question of when the strict topology is a Mackey topology was posed by BUCK and has motivated a great deal of research. The result given here is due to CONWAY [36] although it had already been obtained by LE CAM for σ-compact, locally compact spaces (modulo the identification of the strict topology with a mixed topology). The method given here is from COOPER [41]. ZAFARANI [159] has also simplified Conway's rather complicated proof.

For results of the type mentioned in 1.23, see HOFMANN-JØRGENSEN [79].

II.2. The theory of 2.1 - 2.5 is an attempt to extend the Gelfand-Naimark duality theory for compact spaces to completely regular spaces. Unfortunately, a perfect duality cannot be obtained in this context (see Appendix). The idea of a perfect algebra is borrowed from APOSTOL [1]. For detailed accounts of the Stone-Čech compactification and other extensions of topological spaces, see GILLMAN and JERISON [58], WALKER [145] and WEIR [148]. For a functional-analytic approach, see BUCHWALTER [21].

II.3. One of the most important justifications for the study of strict topologies is the fact that the dual space is the space of bounded Radon (or tight) measures. This fact was proven by BUCK [27] for locally compact spaces. For completely regular spaces, see FREMLIN et al. [50], GILES [54], GULICK [66], HIRSCHFELD [77], HOFMANN-JØRGENSEN [79] and SENTILLES [132]. For a detailed treatment of Radon measures on completely regular spaces, see BOURBAKI [14] and SCHWARTZ [125]. The results and proofs of 3.9 - 3.16 are from FREMLIN et al. [50].

II.4. Strict topologies on spaces of vector-valued continuous functions were studied already by BUCK in [27]. See also FONTENOT [49] and WELLS [150].

The exponential law for non-compact spaces is discussed by SEMADENI [127]. See also NOBLE [109].

The β-dual of $C^{\infty}(X;E)$ can be identified with a certain space of E'-valued measures on X. See KATSARAS [83] - [85]. 4.7 is a result of BUCK [28].

For a comprehensive study of very general locally convex spaces of vector-valued continuous functions, see BIERSTEDT [6] and [7].

II.5. For detailed studies of σ-additive and τ-additive measures, see VARADARAJAN [144] and KNOWLES [92]. The topologies β_{σ} and β_{τ} were introduced in the form given here by SENTILLES [132]. Equivalent topologies were defined by FREMLIN et al. [50] using order theoretical methods. The general topology β_{L} was studied by MOSIMAN [104]. Similar methods have been used by BINZ to define convergence structures on C(X) (see [9] and [10]). β_{∞} was introduced by WHEELER [152]. The corresponding dual space has been studied in various contexts by DUDLEY [46], LEGER and SOURY [95] and ROME [119]. The characterisation of β_{∞} as a β_{L} is due to MOSIMAN [104]. For 5.10 see VARADARAJAN [144]. 5.11 and 5.12 are due to KNOWLES. Measure compactness and strong measure compactness have been studied by KIRK [87], [88] and [89], MORAN [101], [102], [103] and, from a functional analytic view point, by SENTILLES [132] and MOSIMAN and WHEELER [105].

We mention that the topics of II.5 have been covered in the following survey articles: BUCHWALTER [23], COLLINS [31], GULICK [67] and HIRSCHFELD [78].

A detailed study of topological measure theory can be found in TOPSØE [142]. Non-bounded measures are studied by SONDERMANN [135]. For functional-analytic treatment of vector-valued measures see BUCCHIONI [18], DEBIEVE [43].

REFERENCES FOR CHAPTER II.

[1] C. APOSTOL b*-algebras and their representations, Jour. Lond. Math. Soc. (2) 3 (1971) 30-38.

[2] A. BADRIKIAN Seminaire sur les fonctions aléatoires linéaires et les mesures cylindriques (Springer Lecture Notes 139, 1970).

[3] A.G.A.G. BARBIKER Uniform continuity of measures on completely regular spaces, Jour. Lond. Math. Soc. (2) 5 (1972) 451-458.

[4] J. BERRUYER, B. IVOL Espaces de mesures et compactologies, Publ. Dép. Math. Lyon 9-1 (1972) 1-35.

[5] A. BEURLING Une théorème sur les fonctions bornées et uniformement continues sur l'axe reel, Acta Math. 77 (1945) 127-136.

[6] K. BIERSTEDT Gewichtete Räume stetiger vektorwertiger Funktionen und das injektive Tensorprodukt I,II, J. f.d. reine und ang. Math. 259 (1973) 186-210 and 260 (1973) 133-146.

[7] Injektive Tensorprodukte und Slice-Produkte gewichteter Räume stetiger Funktionen, J. f.d. reine und angew. Math. 266 (1974) 121-131.

[8] K. BIERSTEDT, R. MEISE Lokalkonvexe Unterräume in topologischen Vektorräumen und das ε-Produkt, Man. Math. 8 (1973) 143-172.

[9] E. BINZ Continuous convergence on C(X) (Springer Lecture Notes 469 - Berlin, 1975).

[10] E. BINZ, K. FELDMAN On a Marinescu structure on C(X), Comm. Math. Helv. 46 (1971) 436-450.

[11] E. BINZ, K. KUTZLER Über metrische Räume und $C_c(X)$, Ann. Scuola norm. sup. Pisa (III) 26 (1972) 197-223.

[12] M. BLANCHARD, M. JOURLIN La topologie de la convergence bornée sur les algèbres de fonctions continues, Publ. Dép. Math. Lyon 6-2 (1969) 85-96.

[13] N. BOURBAKI Eléments de Mathématique XIII - Intégratio Ch. 1-4 (Paris, 1965).

[14] Eléments de Mathématique XXXV - Intégratio Ch. 9 (Paris, 1969).

[15] General topology - Part 2 (Paris, 1966).

[16] R. BROWN Ten topologies on X × Y, Quart. J. Math. 1 (1963) 303-319.

[17] Function spaces and product topologies, Quart. J. Math. 15 (1964) 238-250.

[18] D. BUCCHIONI Mesures vectorielles et partitions de l'unité, Publ. Dép. Math. Lyon 12-3 (1975) 51-90.

[19] D. BUCCHIONI, A. GOLDMAN Sur certains espaces de formes linéaires liés aux mesures vectorielles, A.. Inst. Fourier 26 (1976) 173-204.

[20] H. BUCHWALTER Espaces de Banach et dualité, Publ. Dép. Math. Lyon 3-2 (1966) 2-61.

[21] Topologie et compactologie, Publ. Dép. Mat Lyon 6-2 (1969) 1-74.

[22] Parties bornées d'un espace topologique complètement régulier (sem. Choquet 1969/ 1970).

[23] H. BUCHWALTER Fonctions continues et mesures sur un espace complètement régulier, Summer School on topological vector spaces (Lecture Notes in Math. 331 (1973) 183-202).

[24] Quelques curieuses topologies sur $M_\sigma(T)$ et $M_\beta(T)$, Ann. Inst. Fourier.

[25] H. BUCHWALTER, J. SCHMETS Sur quelques propriétés de l'espace $C_s(T)$, J. Math. pure appl. 52 (1973) 337-352.

[26] R.C. BUCK Operator algebras and dual spaces, Proc. Amer. Math. Soc. 3 (1952) 681-687.

[27] Bounded continuous functions on a locally compact space, Mich. Math. J. 5 (1958) 95-104.

[28] Approximation properties of vector valued functions, MRC Technical Report Nr. 1359, 1973.

[29] A.K. CHILANA The space of bounded sequences with the mixed topology, Pac. J. Math. 48 (1973) 29-33.

[30] H.S. COLLINS On the space $\ell^\infty(S)$, with the strict topology, Math. Z. 106 (1968) 361-373.

[31] Strict, weighted, and mixed topologies, and applications, Advances in Math. 19 (1976) 207-237.

[32] H.S. COLLINS, J.R. DORROH Remarks on certain function spaces, Math. Ann. 176 (1968) 157-168.

[33] H.S. COLLINS, R.A. FONTENOT Approximate identities and the strict topology, Pac. J. Math. 43 (1972) 63-79

[34] W.W. COMFORT, S. NEGREPONTIS The theory of ultrafilters (Berlin, 1974).

[35] Continuous pseudometrics (New York, 1975).

[36] J.B. CONWAY The strict topology and compactness in the space of measures II, Trans. Amer. Math. Soc. 126 (1967) 474-486.

[37] A theorem on sequential convergence of measures and some applications, Pac. J. Math. 28 (1969) 53-60.

[38] On extending interpolating sets in the Stone Čech compactification, Duke Math. J. 36 (1969) 753-759.

[39] J.B. COOPER The strict topology and spaces with mixed topology, Proc. Amer. Math. Soc. 30 (1971) 583-592.

[40] La loi exponentielle et la topologie mixte, C.R. Acad. Sci. Paris, A 279 (1974) 919-920.

[41] The Mackey topology as a mixed topology, Proc. Amer. Math. Soc. 53 (1975) 107-112.

[42] J.A. CRENSHAW Positive linear operators for the strict topologies, Proc. Amer. Math. Soc. 46 1974) 79-85.

[43] C. DEBIEVE Mesures vectorielles sur les espaces topologiques I,II, Ann. Soc. Sci. Bruxelle 88 (1974) 37-54 and 157-168.

[44] W.E. DIETRICH A note on the ideal structure of C(X), Proc. Amer. Math. Soc. 23 (1969) 174-178.

[45] J.R. DORROH The localization of the strict topology via bounded sets, Proc. Amer. Math. Soc. 20 (1969) 413-414.

[46] R.M. DUDLEY Convergence of Baire measures, Studia Math. 27 (1966) 252-268.

[47] A. ETCHEBERRY Isomorphisms of spaces of bounded continuous functions, Studia Math. 53 (1975) 103-127.

[48] W.A. FELDMAN A characterisation of the topology of compact convergence on C(X), Pac. J. Math. 51 (1974) 109-119.

[49] R.A. FONTENTOT Strict topologies for vector valued functions, Can. J. Math. 26 (1974) 841-853.

[50] D.H. FREMLIN, D.J.H. GARLING, R.G. HAYDON Bounded measures on topological spaces, Proc. Lond. Math. Soc. 25 (1972) 115-136.

[51] P. GÄNSSLER Compactness and sequential compactness in spaces of measures, Z. Wahrscheinlichkeitsth. 17 (1971) 124-146.

[52] A convergence theorem for measures in regular Hausdorff spaces, Math. Scand. 29 (1971) 237-244.

[53] P. GERARD Un critère de compacité dans l'espace $M_t^+(E)$, Bull. Soc. Roy. Liège 42 (1973) 179-182.

[54] R. GILES A generalization of the strict topology, Trans. Amer. Math. Soc. 161 (1971) 467-474.

[55] L. GILLMAN, M. JERISON Rings of continuous functions (Princeton, 1960).

[56] I. GLICKSBERG Bishop's generalized Stone-Weierstraß theorem for the strict topology, Proc. Amer. Math. Soc. 14 (1963) 329-333.

[57] A. GOLDMAN Prémesures et mesures sur les espaces compactologiques, Publ. Dép. Math. Lyon 9-1 (1972) 61-86.

[58] E. GRANIRER On Baire measures on D-topological spaces Fund. Math. 60 (1967) 1-22.

[59] W. GRÖMIG On a weakly closed subset of the space of τ-smooth measures, Proc. Amer. Math. Soc. 43 (1974) 397-401.

[60] A. GROTHENDIECK Critères de compacité dans les espaces fonctionelles généraux, Amer. J. Math. 192 (1952) 168-186.

[61] Sur les applications linéaires faiblement compactes d'espaces du type C(K), Can. J. Math. 5 (1953) 129-173.

[62] Topological vector spaces (New York, 1973

[63] D. GULICK The σ-compact-open topology and its relatives, Math. Scand. 30 (1972) 159-176.

[64] Duality theory for the topology of simple convergence, J. Math. pur. et appl. 52 (1973) 453-472.

[65] Domination in analysis, Publ. Dép. Math. Lyon 10-3 (1973) 35-64.

[66] Duality theory for the strict topology, Studia Math. 49 (1974) 195-208.

[67] Duality theory for spaces of continuous functions, Technical Report - Univ. Maryland, 1974.

[68] D. GULICK, J. SCHMETS Separability and semi-norm separability for spaces of bounded continuous functions, Bull. Soc. Roy. Sc. Liège 41 (1972) 254-260.

[69] F. GULICK, D. GULICK Boundedness for spaces of continuous functions, Technical Report - Univ. Maryland, 1974.

[70] R. HAYDON Sur les espaces M(T) et $M^\infty(T)$, C.T. Acad. Sci. Paris, 275 (1972) A989-991.

[71] Sur un problème de H. Buchwalter, C.R. Acad. Sci. Paris 275 (1972) A1077-1079.

[72] Compactness in $C_s(T)$ and applications, Publ. Dép. Math. Lyon 9-1 (1972) 105-113.

[73] Three examples in the theory of spaces of continuous functions, Publ. Dép. Math. Lyon 9-1 (1972) 99-103.

[74] On compactness in spaces of measures and measure compact spaces, Proc. Lond. Math. Soc. (3) 29 (1974) 1-16.

[75] On the Stone-Weierstraß theorem for the strict and superstrict topologies, Proc. Amer. Math. Soc. 59 (1976) 273-278.

[76] C.S. HERZ The spectral theory of bounded functions, Trans. Amer. Math. Soc. 94 (1960) 181-232.

[77] R.A. HIRSCHFELD On measures in completely regular spaces, Bull. Soc. Math. Belg. 24 (1972) 374-386.

[78] RIOTS, Nieuw Arch. v. Wiskunde (3) 22 (1974) 1-43.

[79] J. HOFMANN-JØRGENSEN A generalisation of the strict topology, Math. Sc. 30 (1972) 313-323.

[80] Weak compactness and tightness of subsets of M(X), Math. Sc. 31 (1972) 127-150.

[81] S. KAPLAN On weak compactness in the space of Radon measures, Jour. Func. Anal. 5 (1970) 259-298.

[82] A.K. KATSARAS On the strict topology in the locally convex setting, Math. Ann. 216 (1975) 105-111.

[83] Vector valued measures and strict topologies, Annali di Mat. 106 (1975) 259-272.

[84] Spaces of vector measures, Trans. Amer. Math. Soc. 206 (1975) 313-328.

[85] On the space C(X,E) with the topology of simple convergence, Math. Ann. 223 (1976) 105-117.

[86] S.S. KHURANA Strict topology on paracompact locally compact spaces, Can. J. Math. 29 (1977) 216-219.

[87] R.B. KIRK Measures on topological spaces and B-compactness, Indag. Math. 31 (1969) 172-183.

[88] Locally compact, B-compact spaces, Indag. Math. 31 (1969) 333-344.

[89] Kolmogorov type consistency theorems for products of locally compact, B-compact spaces, Indag. Math. 32 (1970) 77-81.

[90] Algebras of bounded real-valued functions I,II, Indag. Math. 34 (1972) 443-451 and 452-263.

[91] Complete topologies on spaces of Baire measures, Trans. Amer. Math. Soc. 184 (1973) 1-29.

[92] J.D. KNOWLES Measures on topological spaces, Proc. Lond Math. Soc. (3) 17 (1967) 139-156.

[93] J. GIL DE LAMADRID Measures and tensors I,II, Trans. Amer. Math. Soc. 114 (1965) 98-121 and Can. Jour. Math. 18 (1966) 762-793.

[94] L. LE CAM Convergence in distributions of stochastic processes, Univ. Cal. Publ. Statistics 11 (1975) 207-236.

[95] C, LEGER, P. SOURY Le convexe topologique des probabilitès sur un espace topologique, J. Math. Pure Appl. 50 (1971) 363-425.

[96] R. MACDOWELL Banach spaces and algebras of continuous functions, Proc. Amer. Math. Soc. 6 (1955) 67-78.

[97] S. MACHADO, J.B. PROLLA An introduction to Nachbin spaces, Rend. Circ. Mat. Palermo II 31 (1972) 119-139.

[98] J. MAŘIK Les fonctionelles sur l'ensemble des fonctions continues bornées, définies dans un ensemble topologique, Studia Math. 16 (1957) 86-94.

[99] E. MICHAEL Locally multiplicatively convex topological algebras, Mem. Amer. Math. Soc. 11 (1952).

[100] On k spaces, k_R spaces and k(X), Pac. J. Math. 47 (1973) 487-498.

[101] W. MORAN The additivity of measures on completely regular spaces, Jour. Lond. Math. Soc. 43 (1968) 633-639.

[102] Measures and mappings in topological spaces, Proc. Lond. Math. Soc. (3) 19 (1969) 493-508.

[103] Measures on metacompact spaces, Proc. Lond. Math. Soc. (3) 20 (1970) 507-524.

[104] S. MOSIMAN Strict topologies and the ordered vector space C(S) (Preprint).

[105] S. MOSIMAN, R.F. WHEELER The strict topology in a completely regular setting: relations to topological measure theory, Can. J. Math. 24 (1972) 873-890.

[106] L. NACHBIN On the priority of algebras of continuous functions in weighted approximation, Symp. Math. XVII (1976) 169-183.

[107] L. NACHBIN, S. MACHADO, J.B. PROLLA Weighted approximati vector fibrations, and algebras of operators, J. Math. pures and appl. 50 (1971) 299-323.

[108] L.D. NEL Theorems of Stone-Weierstraß type for non compact spaces, Math. Z. 104 (1968) 226-230.

[109] N. NOBLE Ascoli theorems and the exponential map, Trans. Amer. Math. Soc. 143 (1969) 393-411.

[110] K. NOUREDDINE Topologies strictes sur C(T) et $C^{\infty}(T)$, C.R. Acad. Sc. Paris 280 (1975) A1129-113

[111] K. NOUREDDINE, W. HABRE Topologies P-strictes, Séminaire d'analyse fonctionelle, Fac. Sc. Univ. Libanaise, 1974-1975.

[112] J. GARAY DE PABLO Integracion en espacios topologicos, Zaragoza 22 (1967) 83-124.

[113] D. POLLARD Compact sets of tight measures, Studia Math. 56 (1976) 63-67.

[114] D. POLLARD, F. TOPSØE A unified approach to Riesz type representation theorems, Studia Math. 54 (1975) 173-190.

[115] D. PREISS Metric spaces in which Prohorov's theorem is not valid, Z. Wahrsch. verw. Geb. 27 (1973) 109-116.

[116] J.B. PROLLA Bishop's generalized Stone-Weierstraß theorem for weighted spaces, Math. Ann. 191 (1971) 283-289.

[117] J.B. PROLLA Weighted approximation and slice products of modules of continuous functions, Ann. Sc. Norm. Sup. Pisa (3) 26 (1972) 563-571.

[118] R. PUPIER Méthodes fonctorielles en topologie générale (Thèse Science Mathématiques, Lyon, 1971).

[119] M ROME L'espace $M^{\infty}(T)$, Publ. Dép. Math. Lyon 9-1 (1972) 37-60.

[120] A.C.M. VAN ROOIJ Tight functionals and the strict topology, Kyungpook Math. J. 7 (1967) 41-43.

[121] J. SCHMETS Separability for seminorms on spaces of bounded continuous functions, Jour. Lond. Math. Soc. (2) 11 (1975) 245-248.

[122] Espaces associès à un espace linéaire à seminormes: applications aux espaces de fonctions continues, Sem. Liège, 1972-1973.

[123] Espaces de fonctions continues (Springer Lecture Notes 519, Berlin 1976).

[124] J. SCHMETS, J. ZAFARANI Topologie strict faible et mesures discrètes, Bull. Soc. Roy. Sci. Liège 43 (1974) 405-418.

[125] L. SCHWARTZ Radon measures (Oxford, 1973).

[126] Z. SEMADENI Inverse limits of compact spaces and direct limits of spaces of continuous functions, Studia Math. 31 (1968) 373-382.

[127] Z. SEMADENI Banach spaces of continuous functions, I (Warsaw, 1971).

[128] F.D. SENTILLES Compactness and convergence in the space of measures, III, J. Math. 13 (1969) 761-768.

[129] Compact and weakly compact operators on $C(S)_\beta$, III. J. Math. 13 (1969) 769-776.

[130] The strict topology on bounded sets, Pac. J. Math. 34 (1970) 529-540.

[131] Conditions for equality of the Mackey and strict topologies, Bull. Amer. Math. Soc. 76 (1970) 107-112.

[132] Bounded continuous functions on a completely regular space, Trans. Amer. Math. Soc. 168 (1972) 311-336.

[133] F.D. SENTILLES, R.F. WHEELER Additivity of functionals and the strict tolology (Preprint).

[134] Linear functionals and partitions of unity in $C_b(X)$, Duke J. Math. 41 (1974) 483-496

[135] D. SONDERMANN Masse auf lokalbeschränkten Räumen, Ann. Inst. Math. Fourier 19 (1970) 33-113.

[136] P.D. STRATIGOS Relative compactness in the vague topology Boll. Un. Mat. Ital. (4) 8 (1973) 198-202

[137] W.H. SUMMERS Weighted spaces and weighted approximation Séminaire d'analyse moderne - Sherbrooke, 1970.

[138] The general complex bounded case of the strict weighted approximation problem, Math. Ann. 192 (1971) 90-98.

[139] W.H. SUMMERS Separability in the strict and substrict topology, Proc. Amer. Math. Soc. 35 (1970) 507-514.

[140] C. TODD Stone-Weierstraß theorems for the strict topologies, Proc. Amer. Math. Soc. 16 (1965) 654-659.

[141] A. TONG, D. WILKEN Weak and norm sequential convergence in M(S), J. Aust. Math. Soc. 15 (1973) 1-6.

[142] F. TOPSØE Topology and measure (Berlin, 1970).

[143] Compactness and tightness in a space of measures with the topology of weak convergence, Math. Scand. 34 (1974) 187-210.

[144] V.S. VARADARAJAN Measures on topological spaces , Amer. Math. Soc. Transl. (2) 48 (1965) 161-228.

[145] R.C. WALKER The Stone-Čech compactification (Berlin, 1974).

[146] S. WARNER The topology of compact convergence on continuous function spaces, Duke Math. J. 25 (1958) 265-282.

[147] J.H. WEBB The strict topology is mixed.

[148] M.D. WEIR Hewitt-Nachbin spaces (Amsterdam, 1975).

[149] B.W. WELLS Jr. Weak compactness of measures, Proc. Amer. Math. Soc. 20 (1969) 124-130.

[150] J. WELLS Bounded continuous vector valued functions on a locally compact space, Mich. Math. J. 12 (1965) 119-126.

[151] R.F. WHEELER The equicontinuous weak * topology and semi-reflexivity, Studia Math. 41 (1972) 243-256.

[152] R.F. WHEELER The strict topology, separable measures, and paracompactness, Pac. J. Math. 47 (1973) 287-302.

[153] The strict topology for P-spaces, Proc. Amer. Math. Soc. 41 (1973) 466-472.

[154] A locally compact non paracompact space for which the strict topology is Mackey, Proc. Amer. Math. Soc. 51 (1975) 86-90.

[155] Well-behaved and totally bounded approximate identities for $C_O(X)$, Pac. J. Math. 65 (1974) 261-269.

[156] The Mackey problem for the compact open topology, Trans. Amer. Math. Soc. 222 (1976) 255-265.

[157] S. WILLARD General topology (Addison Wesley, 1970).

[158] R.G. WOODS Topological extension properties, Trans. Amer. Math. Soc. 210 (1975) 365-385.

[159] J. ZAFARANI A sufficient condition for the strict topology to be Mackey, Bull. Soc. Roy. Sc. Liège 44 (1975) 569-571.

CHAPTER III - SPACES OF BOUNDED, MEASURABLE FUNCTIONS

Introduction: In this chapter, we consider the L^∞-space associated with a positive Radon measure on a locally compact space. We define two supplementary topologies which give it the structure of a Saks space. In 1.1 - 1.5 we discuss the basic properties of the corresponding mixed topologies β_σ and β_1. In 1.6 - 1.8 we show that both mixed topologies are topologies of the dual pair (L^∞, L^1) and that β_1 is, in fact, the Mackey topology. The latter result is equivalent to the DUNFORD-PETTIS theorem (on relatively weakly compact subsets of L^1) but this formulation sheds new light on the result and the theory of Chapter I allows us to deduce some important consequences. In 1.18 we use it to prove that C(K) has the Dunford-Pettis property (following GROTHENDIECK). The assumption that μ is a Radon measure is actually unnecessary and we close the section with some remarks on how the results can be extended to L^∞-spaces associated with abstract measures.

In the second section we consider β_1-continuous linear operators on L^∞ and show that they are induced by vector-valued measures.

In section 3, we consider the theory of measurable functions with values in a Saks space. In order to cover this theme properly, it is necessary to use much heavier machinery than we are prepared to use in this book so that this section is no more than a sketch of a possible theory. However, several facts indicate that the development of such a theory would be worthwhile. For example, it would allow a synthesis of various concepts of vector valued integration (Bochner integral, Pettis and Gelfand integrals and the lower star integral spaces introduced by L. SCHWARTZ). We end the Chapter with some remarks on the Radon-Nikodym property for Banach spaces.

III.1. THE MIXED TOPOLOGIES

In this Chapter we study spaces of bounded measurable functions on a measure space. We begin with some remarks on the simplest case - that of a discrete measure space i.e. the space $\ell^\infty(S)$ for some indexing set S. We consider this as a Saks space with the supremum norm and the topology τ_p of pointwise convergence (see I.2.B). Note that

$$(\ell^\infty(S), \|\ \|, \tau_p) = (C^\infty(S), \|\ \|, \tau_K)$$

where we regard S as a topological space with the discrete topology. We also have

$$(\ell^\infty(S), \|\ \|, \tau_p) = S \prod_{\alpha \varepsilon S} E_\alpha$$

where each E is the canonical one-dimensional Saks space. We now list some properties of $\ell^\infty(S)$ and the associated mixed topology β which follows immediately from the theory of the earlier chapters:

1) $(\ell^\infty(S), \beta)$ is a complete, semi-Montel locally convex algebra in which multiplication is jointly continuous (II.2.1)

2) the dual of $(\ell^\infty(S),\beta)$ is canonically identifiable with $\ell^1(S)$ and $\beta = \tau_c(\ell^\infty,\ell^1) = \tau(\ell^\infty,\ell^1)$ in particular, $(\ell^\infty(S),\beta)$ is a Mackey space (I.4.13 and I.4.14);

3) $(\ell^\infty(S),\beta)$ has the Banach-Steinhaus property, is B_r-complete and satisfies a closed graph theorem with range space a separable Fréchet space (I.4.16);

4) if S is countable, the unit ball of $\ell^\infty(S)$ is β-metrisable and so a subset A of $\ell^\infty(S)$ is β-closed if and only if it is sequentially closed;

5) $(\ell^\infty(S),\beta)$ is separable if and only if $\text{Card}(S) \le \text{Card}(\mathbb{R})$ (II.1.17).

Now let M be a locally compact space, μ a positive Radon measure on M. $L^p(M;\mu)$ $(1 \le p \le \infty)$ denotes the space of measurable, complex functions x and M so that

$$\|x\|_p := \{ \int_M |x(t)|^p d\mu \}^{1/p} \qquad (1 \le p < \infty)$$

$$\|x\|_\infty := \inf \{ M_1 > 0 : |x(t)| \le M_1 \text{ locally } \mu\text{-almost everywhere} \}$$

respectively is finite where as usual functions x and y are identified if they agree locally μ-almost everywhere (i.e. if, for each compact $K \subseteq M$, $\mu(\{t \in K : x(t) \neq y(t)\}) = 0$). Then $(L^p, \|\ \|_p)$ is a Banach space for each p. (L^∞, L^1) is a dual pair under the bilinear form

$$(x,y) \longmapsto \int_M x(t)y(t)d\mu$$

and under this duality, L^∞ is identifiable with the dual of the Banach space $(L^1, \| \|_1)$. Hence we can define on L^∞ the topology $\sigma := \sigma(L^\infty, L^1)$ and

$$(L^\infty, \| \|_\infty, \sigma)$$

is a Saks space. We denote by β_σ the mixed topology $\gamma[\| \|_\infty, \sigma]$. We resume its properties in the following Proposition:

1.1. Proposition: 1) β_σ is a topology of the dual pair (L^∞, L^1) and is, in fact, the topology $\tau_c(L^\infty, L^1)$ of uniform convergence on the compact subsets of $(L^1, \| \|_1)$;

2) β_σ has the same convergent sequences and compact subsets as $\sigma(L^\infty, L^1)$;

3) (L^∞, β_σ) is a complete, semi-Montel space;

4) if M is σ-compact and metrisable, then $B_{\| \|_\infty}$ is β_σ-metrisable and so a linear operator from L^∞ into a locally convex space F is β_σ-continuous if and only if it is sequentially continuous.

Proof: We remark only that 4) follows from the fact that under these conditions, $(L^1, \| \|_1)$ is separable.

We shall now introduce a more interesting mixed topology on L^∞. We require two results on measurable functions which we now state. First recall the following definition:

a subset H of $L^p(M;\mu)$ is p-equi-integrable $(1 \leq p < \infty)$ if the following conditions are satisfied:

a) for every $\varepsilon > 0$ there is a $\delta > 0$ so that if A is an integrable subset of M with $\mu(A) \leq \delta$ then

$$\int_A |x(t)|^p \, d\mu \leq \varepsilon \qquad \text{for each } x \in H;$$

b) for each $\varepsilon > 0$ there is a compact subset K of M so that

$$\int_{M\setminus K} |x(t)|^p \, d\mu \leq \varepsilon \qquad \text{for each } x \in H.$$

We remark that if M is compact (so that $L^\infty(M;\mu) \subseteq L^p(M;\mu)$) then the unit ball of $(L^\infty, \| \|_\infty)$ is p-equi-integrable for $1 \leq p < \infty$.

1.2. Proposition: If H is a p-equi-integrable subset of $L^p(M;\mu)$ then on H the uniformity induced by the norm $\| \|_p$ coincides with the uniformity of convergence in measure on compact sets.

Proof: If $\varepsilon > 0$ is given, choose $\delta > 0$ and K as in a) and b). We show that if x,y in H are such that

$$|x(t) - y(t)| \leq \varepsilon^{1/p}(\mu(K))^{-1/p}$$

for $t \in K \setminus A$ where $\mu(A) \leq \delta$, then $\|x - y\|_p^p \leq (2^{p+1} + 1)\varepsilon$

For we can estimate as follows:

$$\left(\int_{M\setminus K} |x(t)-y(t)|^p d\mu\right)^{1/p} \leq$$

$$\leq \left(\int_{M\setminus K} |x(t)|^p d\mu\right)^{1/p} + \left(\int_{M\setminus K} |y(t)|^p d\mu\right)^{1/p}$$

$$\leq 2\,\varepsilon^{1/p}$$

Similarly

$$\left(\int_A |x(t)-y(t)|^p d\mu\right)^{1/p} \leq 2\,\varepsilon^{1/p}$$

Hence

$$\|x - y\|_p^p \leq \left(\int_{M\setminus K} + \int_A + \int_{K\setminus A} \right) |x(t)-y(t)|^p d\mu$$

$$\leq 2^p\varepsilon + 2^p\varepsilon + \frac{\varepsilon}{\mu(K)} (\mu(K \setminus A)) \leq (2^{p+1} + 1)\varepsilon$$

1.3. Corollary: If M is compact, then the uniformity induced on $B_{\|\ \|^\infty}$, the unit ball of $(L^\infty, \|\ \|_\infty)$, by the norm $\|\ \|_p$ $(1 \leq p < \infty)$ coincides with that of convergence in measure.

1.4. Proposition: Let $\{x_K : K \in K(M)\}$ be a family where x_K is a measurable, complex-valued function on K so that if $K \subseteq K_1$ then $x_{K_1} = x_K$ almost everywhere on K. Then there is a measurable function x on M so that $x = x_K$ almost everywhere on K for each $K \in K(M)$.

Proof: Consider the family of pairs (x,U) where U is an open subset of M and x is a measurable function on U so that $x = x_K$ almost everywhere on K $(K \subseteq U)$. We order this family by putting

$$(x,U) \leq (y,V)$$

if $U \subseteq V$ and y is an extension of x. Its is clear that this ordering is inductive and so there exists a maximal element (x,W) by Zorn's Lemma. Then W = M by maximality and so x is the required function.

We now consider on $L^\infty(M;\mu)$ the locally convex topology τ^p_{loc} $(1 \leq p < \infty)$ defined by the seminorms

$$\| \|_K^p : x \longmapsto (\int_K |x(t)|^p \, d\mu)^{1/p}$$

as K runs through $K(M)$. Then $(L^\infty, \| \|_\infty, \tau^p_{loc})$ is a complete Saks space. To prove this, we consider first the case where M is compact (so that τ^p_{loc} is just the norm topology induces by $\| \|_p$). If (x_n) is a $\| \|_p$-Cauchy sequence in $B_{\| \|_\infty}$, then it has an L^p-limit x. We can extract a subsequence of (x_n) which converges pointwise μ-almost everywhere to x (cf. BOURBAKI [2], IV.3.4., Théorème 3) and so $x \in B_{\| \|}$.

Now in the general case (i.e. where M is not necessarily compact). $\{L^\infty(K) : K \in K(M)\}$ forms a projective system of Saks spaces (under the restriction mappings) and, by 1.4, $L^\infty(M;\mu)$ is its Saks space projective limit $S\text{-}\varprojlim \{ L^\infty(K)\}$ (for an element of the projective limit is just a thread $\{x_K\}$ of the type described in the Proposition and so defines an element of $L^\infty(M)$).

Hence we can define the mixed topologies

$$\beta_p := \gamma \left[\| \|_\infty, \tau^p_{loc} \right]$$

Now by 1.3, $\tau^p_{loc} = \tau^q_{loc}$ on $B_{\| \|_\infty}$ and so these topologies coincide. We shall denote this topology by β_1.

The following result follows directly from the theory of Chapter I.

1.5. <u>Proposition</u>: 1) (L^∞, β_1) is complete;

2) a sequence (x_n) in L^∞ is β_1-convergent to zero if and only if it is $\| \|_\infty$-bounded and convergent to zero in some τ^p_{loc} or even in measure on compact sets;

3) a linear operator from L^∞ into a locally convex space F is β_1-continuous if and only if its restriction to $B_{\|\ \|_\infty}$ is τ^p_{loc}-continuous for some $p \in [1,\infty[$;

4) if M is σ-compact, then $B_{\|\ \|_\infty}$ is β_1-metrisable and so a linear operator from L^∞ into F is β_1-continuous if and only if it is sequentially continuous.

1.6. <u>Proposition</u>: The dual of (L^∞,β_1) is identifiable with $(L^1,\|\ \|_1)$ under the bilinear form

$$(x,y) \longmapsto \int_M x(t)y(t)d\mu$$

<u>Proof</u>: Since (L^∞,τ^1_{loc}) is a dense subspace of the locally convex projective limit of the spaces $\{L^1(K) : K \in \mathcal{K}(M)\}$ (i.e. the space of locally integrable functions), we can identify its dual with the subspace of $L^1(M)$ consisting of those functions in L^∞ which have compact support. The closure of the latter space in the L^1-norm is clearly L^1 and this is the required dual space by I.1.18.(ii).

We now come to the main result of this section, namely, that (L^∞,β_1) is a Mackey space.

1.7. <u>Lemma</u>: Let H be a subset of $L^1(M;\mu)$ that is not 1-equi-integrable. Then there exists an $\alpha > 0$, a sequence (x_n) in H and a disjoint sequence (A_n) of measurable sets so that $|\int_{A_n} x_n\, d\mu| \geq \alpha$ for each n.

Proof: Firstly, we note that if $x \in L^1$ and $\varepsilon > 0$ then there is a $\delta > 0$ so that if $\mu(A) < \delta$ then $\int_A |x| d\mu \leq \varepsilon$. For if $M_n := \{t \in M : |x(t)| \geq n\}$ then $\bigcap M_n = \phi$ and so, by Lebesgue's theorem on dominated convergence, $\int_{M_n} |x| d\mu \longrightarrow 0$. i.e. there exists an $n \in \mathbb{N}$ so that $\int_{M_n} |x| d\mu \leq \varepsilon/2$.

Let $\delta := \varepsilon(2n)^{-1}$. Then if $\mu(A) < \delta$ we have

$$\int_A |x| d\mu \leq \left(\int_{A \setminus M_n} + \int_{M_n} \right) |x| d\mu \leq n\mu(A) + \varepsilon/2 \leq \varepsilon.$$

We shall suppose that condition a) in the definition of equi-integrability is not verified (the proof for the case where b) is violated is simpler). Then there is a $\varepsilon > 0$ so that for each $\delta > 0$ there is a measurable set A and an $x \in H$ with $\mu(A) < \delta$ and $\int_A |x| d\mu \geq \varepsilon$. We construct inductively a sequence (B_n) of measurable sets and a sequence (x_n) in H as follows: choose x_1 and B_1 so that $\int_{B_1} |x_1| d\mu \geq \varepsilon$. Now suppose that $x_1, \ldots, x_n$ and $B_1, \ldots, B_n$ are chosen. There is $\delta_n > 0$ so that if A is measurable and $\mu(A) < \delta_n$ we have

$$\int_A |x_k| d\mu < \varepsilon 2^{-n-1} \quad (k = 1, \ldots, n).$$

Then we can choose B_{n+1} and x_{n+1} so that $\mu(B_{n+1}) < \delta_n$ and $\int_{B_n} |x_{n+1}| d\mu \geq \varepsilon$.

Now if we put $A_n := B_n \setminus {}_{k>n} B_k$, one can verify that the A_n are disjoint. Also we have: $\int_{A_n} |x_n| d\mu \geq \varepsilon/2$. The result then follows by a standard argument (reduce to the real-valued case and then consider positive and negative parts).

1.8. Proposition: Let H be a subset of $L^1(M;\mu)$. Then the following conditions are equivalent:

1) H is β_1-equicontinuous;
2) H is relatively $\sigma(L^1, L^\infty)$-compact;
3) H is norm-bounded and 1-equi-integrable.

Proof: 1) $\Longrightarrow$ 2) follows from the ALAOGLU-BOURBAKI theorem.
2) $\Longrightarrow$ 3): it is clear that H is norm-bounded. If it were not 1-equi-integrable, then we could choose (x_n) and (A_n) as in 1.7. Now consider the mapping

$$T : x \longmapsto \left(\int_{A_n} x \, d\mu\right)_{n \in \mathbb{N}}$$

which is continuous from $L^1(M;\mu)$ into $\ell^1(\mathbb{N})$. Then the sequence (Tx_n) is not relatively weakly compact in $\ell^1(\mathbb{N})$ by I.4.14 - contradiction.
3) $\Longrightarrow$ 1): let H be a norm-bounded, equi-integrable subset of L^1. We use the criterium of I.1.22 to show that H is β_1-equicontinuous For $\varepsilon > 0$ we choose $K \in K(M)$ so that $\int_{M \setminus K} |x| d\mu \leq \varepsilon/2$ for each $x \in H$. Choose $n \in \mathbb{N}$ so that $\mu(A) \leq N_1 . n^{-1}$ implies $\int_A |x| d\mu \leq \varepsilon/2$ for each $x \in H$ where $N_1 := \sup\{\|x_1\| : x \in H\}$. Then if $x \in H$, put

$$K' := \{t \in K : |x(t)| \geq n\}.$$

Then $\mu(K') \leq N_1 n^{-1}$ and so, putting $x_1 := x.\chi_{K \setminus K'}$ and $x_2 := x - x_1$, we have

$$x_1 \in H_1 := \{y \in L^1 : \operatorname{supp} y \subseteq K \text{ and } \|y\|_\infty \leq n\}$$

and $\int_M |x_2(t)|d\mu = (\int_{M\setminus K} + \int_K)|x(t)|d\mu \leq \varepsilon.$

Hence $H \subseteq \varepsilon B_{\|\ \|_\infty} + H_1$ and H_1 is τ^1_{loc}-equicontinuous.

1.9. <u>Corollary</u>: (L^∞,β_1) is a Mackey space.

1.10. <u>Corollary</u>: Let T be a continuous linear mapping from $(L^\infty,\sigma(L^\infty,L^1))$ into a locally convex space F with its weak topology $\sigma(F,F')$. Then T is β_1-continuous and so transforms bounded sequences in L^∞ which converge in mean on compact sets into strongly convergent sequences in F.

As an application of 1.10 consider a weakly summable mapping Φ from M into a locally convex space F. That is, for each $f \in F'$, $t \longmapsto f(\Phi(t))$ is in $L^1(M;\mu)$. Then for each $x \in L^\infty$, the integral

$$\int_M f(\Phi(t))x(t)d\mu$$

exists and the mapping $f \longmapsto \int_M f(\Phi(t))x(t)d\mu$ is an element of the algebraic dual $(F')^*$ of F'. The mapping which takes $x \in L^\infty$ into this form is then $\sigma(L^\infty,L^1)$-$\sigma((F')^*,F')$ continuous. Suppose that the range of this mapping lies in F (Φ is then said to be <u>strictly weakly summable or Pettis integrable</u>). Then it is β_1-continuous by 1.10.

We now consider Banach-Steinhaus theorems and closed graph theorems for L^∞. We first show that $(L^\infty,\|\ \|_\infty,\tau^1_{loc})$ satisfies condition Σ_1 of I.4.9.

1.11. Lemma: The Saks space $(L^\infty, \|\ \|_\infty, \tau^1_{loc})$ satisfies condition Σ_1.

Proof: It is convenient to consider real-valued functions. Suppose that $x_0 \in B_{\|\ \|_\infty}$, $\varepsilon > 0$ and $K \in \mathcal{K}(M)$ are given. Let x be an element of $B_{\|\ \|_\infty}$ with $\|x\|^1_K \leq \varepsilon/2$. We shall find a decomposition $x = x_1+x_2$ of x with $x_1, x_2 \in B_{\|\ \|_\infty}$ and $\|x_0-x_1\| \leq \varepsilon$ and $\|x_0-x_2\| \leq \varepsilon$. Denote by M_1 the set where x_0 and x have the same sign.

Then we define x_1 by

$$x_1 := \begin{cases} x + x_0(1-x) & \text{on } M_1 \\ x + x_0(1+x) & \text{on } M \setminus M_1 \end{cases}$$

and $x_2 := x - x_1$.

Then a simple calculation shows that $x_1, x_2 \in B_{\|\ \|_\infty}$ and $\|x_0 - x_1\| \leq \varepsilon$, $\|x_0 - x_2\| \leq \varepsilon$.

In the next results we assume, for simplicity, that M is σ-compact (so that $B_{\|\ \|_\infty}$ is β_1-metrisable). In fact, the results hold without this assumption (see 1.14).

1.12. Proposition: If M is σ-compact, then (L^∞, β_1) has the Banach-Steinhaus property.

Proof: 1.11 and I.4.11.

In particular, this implies the well-known result that $L^1(M;\mu)$ is weakly sequentially complete.

1.13. Proposition: Let M be σ-compact. Then a linear mapping from (L^∞,β_1) into a separable Fréchet space is continuous if and only if it has a closed graph.

1.14. Remark: We sketch briefly the method of extending 1.12 (and hence 1.13) to general M. Since (L^∞,β_1) is a Mackey space, it suffices to show that it has the Banach-Steinhaus property for functionals. Now each functional is represented by an L^1-function and this has σ-compact support. Hence if (f_n) is a pointwise convergent sequence of functionals on L^∞, then there is a σ-compact subset M_o of M so that the corresponding L^1-functions vanish outside of M_o. We can then deduce that the pointwise limit is continuous by applying 1.12 to $L^\infty(M_o;\mu_{M_o})$.

1.15. Remark: We note that it follows from I.4.33.II that (L^∞,β_1) is nuclear if and only if L^∞ is finite dimensional (i.e. if and only if μ has finite support). If μ is atomic (so that (L^∞,β_1) has the form $(\ell^\infty(S),\beta)$ for some set S, then (L^∞,β_1) is semi-Montel and even a Schwartz space. The converse is also true (as DAZORD and JOURLIN have remarked) since if μ is not atomic then, by a result of GROTHENDIECK, L^1 possesses a weakly convergent sequence which is not convergent in the norm. (ℓ^∞,β) is even a universal Schwartz space in the sense that every Schwartz space is a subspace of a product of (ℓ^∞,β)-space.

1.16. Remark: It is perhaps interesting to point out that various ORLICZ-PETTIS theorems on unconditionally summable series (cf. THOMAS [29], § 0) can be obtained from the closed graph theorem (1.13). The point is that if E is a complete locally convex space, there is a bijective correspondence between the set of β_1-continuous linear mappings T from $\ell^\infty(\mathbb{N})$ into E and the unconditionally summable sequences in E, obtained by mapping T into (Te_n) where (e_n) is the standard basis in $\ell^\infty(\mathbb{N})$ (see GROTHENDIECK, Ref. [62] to Chapter II, p.97).

As far as we know, the topology β_1 was first studied (implicitly) by GROTHENDIECK [12] in his investigation of the DUNFORD-PETTIS property, in particular, to show that C(K) has this property. Since we now have sufficient machinery at our disposal to reproduce Grothendieck's proof we do so here. We recall the definition.

1.17. Definition: A locally convex space E has the strict Dunford-Pettis property (abbreviated to SDPP) if every $\sigma(E,E')$-Cauchy sequence is Cauchy for the topology of uniform convergence on the equicontinuous, $\sigma(E',E'')$-compact discs on E' (this is equivalent to the fact that every weakly compact, continuous linear operator from E into a Banach space maps weakly Cauchy sequences into strongly Cauchy sequences). If E is a Banach space, this implies the Dunford-Pettis property for E (i.e. every weakly compact operator from E into a Banach space maps weakly compact sets into relatively compact sets).

1.18. Proposition: Let K be compact. Then the Banach space C(K) has the SDPP.

Proof: Let (x_n) be a weak Cauchy sequence in C(K). Then (x_n) is uniformly bounded and pointwise Cauchy (cf. II.1.21) and so has a bounded, pointwise limit x (which is then measurable for any Radon measure on K). We must show that $(\int_K x_n \, d\mu)$ converges uniformly on weakly compact sets of Radon measures. If this were not the case, we could find a relatively weakly compact sequence (μ_k) of Radon measures on which convergence is not uniform. By a standard construction, there is a positive Radon measure μ on K so that each μ_k is absolutely continuous with respect to μ (take $\mu := \sum_k 2^{-k}|\mu_k|$). Then we can regard $\{\mu_k\}$ as a subset of $L^1(K;\mu)$ and it is relatively weakly compact there. Now by Egoroff's theorem, the sequence (x_n) (regarded as a sequence in $L^\infty(K;\mu)$) converges to x in β_1 and so uniformly on weakly compact subsets of $L^1(K;\mu)$ by 1.9. This gives a contradiction.

1.19. Remark: I. Since a Banach space predual of a space with SDPP has SDPP and the dual of L^1 is a C(K), it can be deduced from this result that every L^1-space has the SDPP.

II. The same proof shows that $(C^\infty(S),\beta)$ has the SDPP for any locally compact space S.

1.20. Remark: We have chosen to develop the theory of L^∞ -spaces in the context of Radon measures on locally compact spaces.

In fact, the theory given here can be developed in the more general setting of abstract measure theory and we indicate briefly how this can be done. Let (M,Σ) be a pair where Σ is a σ-algebra on the set M. A positive regular measure is a σ-additive function μ from Σ into $\mathbb{R}^+ \cup \{\infty\}$ which satisfies the (regularity) condition:

$$\mu(A) = \sup\{\mu(A_o) : A_o \subseteq A, \mu(A_o) < \infty\}$$

for each $A \in \Sigma$.

Then we can develop as usual a theory of integration with respect to μ for complex-valued functions on M and, in particular, we can define the spaces $\mathcal{L}^1(M;\mu)$ and $L^1(M;\mu)$ of absolutely integrable functions respectively of equivalence classes of such functions. $L^1(M;\mu)$ is a Banach space under the norm $\| \|_1 : x \longmapsto \int_M |x| d\mu$. Note that the regularity of μ can be expressed in functional analytic terms as follows: $L^1(M;\mu)$ is the Banach space inductive limit (cf. the introductory remarks to § I.3) of the spectrum

$$\{L^1(A;\mu_A) : A \in \Sigma_o\}$$

where Σ_o denotes the family of sets of Σ with finite μ-measure.

For the time being, we suppose that μ is finite (i.e. $\mu(M) < \infty$). $\mathcal{L}^\infty(M;\mu)$ denotes the space of bounded, measurable complex-valued functions on M and $L^\infty(M;\mu)$ denotes the corresponding quotient space. $L^\infty(M;\mu)$ is a Banach space under the usual norm $\| \|_\infty$.

Then L^1 and L^∞ are in duality with respect to the bilinear form

$$(x,y) \longmapsto \int xy \, d\mu$$

and this establishes L^∞ as the dual of L^1. We can regard L^∞ as a Saks space with the structure $(L^\infty, \| \|_\infty, \| \|_1)$.

We now return to the general case (i.e. where M is not necessarily of finite measure). In order to preserve the duality between L^1 and L^∞ it is necessary to replace the classical L^∞-space by a more complicated one. Recall that $L^1(M;\mu)$ is the Banach space inductive limit of the spaces $\{L^1(A); A \in \Sigma_o\}$. Hence its dual is the Banach space projective limit of the Banach spaces $\{L^\infty(A); A \in \Sigma_o\}$. We define the space $\tilde{L}^\infty(M;\mu)$ to be the set of threads

$$x = \{x_A : A \in \Sigma_o\}$$

where $x_A \in L^\infty(A;\mu_A)$ and

a) $x_{A|A_1} = x_{A_1}$ for $A_1 \subseteq A$ and

b) $\|x\|_\infty := \sup\{\|x_A\|_{L^\infty(A)} : A \in \Sigma_o\}$

Then $\tilde{L}^\infty(M;\mu)$ is a Banach space and can be identified (naturally and isometrically) with the dual of $L^1(M;\mu)$. We note that $\tilde{L}^\infty(M;\mu)$ coincides with the classical space $L^\infty(M;\mu)$ in the following two special cases:

1) where μ is a Radon measure on a locally compact space (this is the content of GODEMENT's Lemma (1.4));

2) M is σ-finite.

Now we can provide $\tilde{L}^{\infty}(M;\mu)$ with a locally convex topology $\tau^1_{\Sigma_o}$ defined by the seminorms

$$x \longmapsto \int_A |x_A| d\mu_A \quad (A \in \Sigma_o)$$

and $(\tilde{L}^{\infty}(M;\mu), \|\ \|_{\infty}, \tau^1_{\Sigma_o})$ is a Saks space - in fact, it is precisely the Saks space projective limit of the system $\{(L^{\infty}(A), \|\ \|_{\infty}, \|\ \|_1); A \in \Sigma_o\}$.

We are thus in the position to define mixed topologies

$$\beta_{\sigma} := \gamma[\|\ \|_{\infty}, \sigma(\tilde{L}^{\infty}, L^1)]$$
$$\beta_1 := \gamma[\|\ \|_{\infty}, \tau^1_{\Sigma_o}]$$

on $\tilde{L}^{\infty}$ and Propositions 1.1, 1.5, 1.6, 1.8, 1.9, 1.10, 1.12. 1.13 can be carried over, *mutatis mutandis*, to these topologies. In particular, β_1 is the Mackey topology of the duality $(\tilde{L}^{\infty}, L^1)$.

III.2. LINEAR OPERATORS AND VECTOR MEASURES

In this section, we characterise the β_1-continuous linear mapping from $L^{\infty}(M)$ into a Banach space. This characterisation displays the close connection between the mixed topology on L^{∞} and vector measures and is one of the bases of applications of mixed topologies to such measures. In this section we use abstract measures rather than Radon measures (cf. the discussion in 1.20) since the latter would involve an unnecessary restriction of generality and would complicate the presentation rather than simplify it.

We recall some definitions and simple results on Banach space valued measures.

2.1. Definition: Let $(M;\Sigma)$ be a measure space (i.e. Σ is a σ-algebra of subsets of M). A measure on $(M;\Sigma)$ with values in a Banach space E is a mapping

$$\mu : \Sigma \longrightarrow E$$

which is additive (i.e. such that $\mu(A \cup B) = \mu(A) + \mu(B)$ for disjoint A and B in Σ).

μ is bounded if (in addition) $\sup\{\|\mu(A)\| : A \in \Sigma\} < \infty$;

σ-additive if $\mu(\bigcup_n A_n) = \sum_n \mu(A_n)$ (convergence in norm) for each disjoint sequence (A_n) in Σ.

Note that the boundedness of μ is equivalent to the condition that $\|\mu\|(M) < \infty$ where $\|\mu\|$, the semi-variation of μ, is the positive set function on Σ defined by

$$\|\mu\|(A) := \sup\{|f \circ \mu|(A) : f \in E', \|f\| \leq 1\} \quad (A \in \Sigma).$$

($|f \circ \mu|$ denotes the variation of the (complex-valued) measure $f \circ \mu$ on Σ).

If $(M;\Sigma)$ is as above, we denote by $B(M;\Sigma)$ the vector space of bounded, Σ-measurable complex-valued functions on M. $B(M;\Sigma)$ is a Banach space with respect to the supremum norm and the subspace of step-functions (i.e. the linear span of the characteristic functions of sets in Σ) is dense.

If $x = \sum_{k=1}^{n} \lambda_k \chi_{A_k}$ $(\lambda_k \in \mathbb{C}, A_k \in \Sigma)$ is a step-function and μ is an E-valued measure, then we can define

$$\int x\, d\mu := \sum_{k=1}^{n} \lambda_k \mu(A_k).$$

Then $x \longmapsto \int x\, d\mu$ is a linear mapping from the space of step-functions into E and we have the estimate:

$$\|\int x\, d\mu\| \leq \|\mu\|(M)\ \|x\|_\infty$$

Hence this mapping extends in a unique manner to a continuous linear mapping from $B(M;\Sigma)$ into E and we continue to denote the action of this mapping by an integral sign. In fact, all norm-continuous linear mappings from $B(M;\Sigma)$ have this form:

2.2. Proposition: The mapping

$$\mu \longmapsto (x \longmapsto \int x\, d\mu)$$

establishes a natural isometry between $L(B(M;\Sigma);E)$ and $M_f(\Sigma;E)$, the space of bounded, E-valued measures on $(M;\Sigma)$ when the latter space is equipped with the norm $\mu \longmapsto \|\mu\|(M)$.

Proof: If $T \in L(B(M;\Sigma);E)$, we can define an E-measure μ_T on M by putting

$$\mu_T(A) := T(\chi_A)$$

for each $A \in \Sigma$. The mapping $T \longmapsto \mu_T$ is an inverse to the mapping defined above. This establishes a vector space isomorphism between the spaces in the statement of the Proposition. That it is an isometry follows from standard estimates.

Now suppose that ν is a positive, regular measure on $(M;\Sigma)$ and μ is an E-measure. We say that μ is <u>ν-continuous</u> if and only if, for each $A \in \Sigma$,

$$\nu(A) = 0 \Longrightarrow \mu(A) = 0.$$

We denote the family of ν-continuous measures by $M_f^{\nu}(\Sigma;E)$.

2.3. <u>Proposition</u>: The isometry

$$L(B(M;\Sigma);E) \cong M_f(\Sigma;E)$$

induces an isometry

$$L(L^{\infty}(M;\nu);E) \cong M_f^{\nu}(\Sigma;E).$$

<u>Proof</u>: The result follows from the simple remark that an operator T lifts from $B(M;\Sigma)$ to the quotient space $L^{\infty}(M;\nu)$ if and only if it vanishes on the characteristic functions of ν-negligible sets.

We now come to the main result of this section - the identification of β_1-continuous linear operators on L^{∞} with vector-valued measures. We say that an E-valued measure μ is <u>ν-regular</u> if for each $A \in \Sigma$

$$\mu(A) = \lim_{\substack{A_0 \subseteq A \\ A_0 \in \Sigma_0}} (\mu(A_0))$$

and we denote by $M_{\sigma}^{\nu}(\Sigma;E)$ the space of all σ-additive measures in $M_f^{\nu}(\Sigma;E)$ which are ν-regular.

2.4. Proposition: The above isometries induce a natural isometry from $L(L^\infty(M;\nu),\beta_1);E)$ onto $M_\sigma^\nu(\Sigma;E)$.

Proof: First we note that if (A_n) is a disjoint sequence in Σ and $A := \bigcup_{n=1}^\infty A_n$ then $\sum_{k=1}^n \chi_{A_k} \longrightarrow \chi_A$ in β_1 (since ν is regular). Hence, if T is β_1-continuous, the induced measure is σ-additive. A similar argument shows that it is regular.

On the other hand, suppose that μ is σ-additive and ν-regular. We must show that the integral operator

$$x \longmapsto \int x \, d\mu$$

is β_1-continuous. Since β_1 is the Mackey topology, we can assume that $E = \mathbb{C}$ and then the result follows from the Radon-Nikodym theorem ([3], Th. 2.2.4).

If we combine 2.4 with the remarks after 1.10 we obtain the fact that a Pettis-integrable function Φ induces a σ-additive measure via the formula $A \longmapsto \int_A \Phi \, d\mu$ (cf. DIESTEL and UHL [8], Ch.I

III.3. VECTOR-VALUED MEASURABLE FUNCTIONS

In this section we consider measurable functions with values in a Saks space. We recall some facts on Banach space valued functions. In order to simplify the notation and avoid technicalities, we shall work with a positive Radon measure on a

σ-compact, locally compact space M. If F is a Banach space, then a function $x : M \longrightarrow F$ is (Lusin) measurable if for each $K \in \mathcal{K}(M)$ and each $\varepsilon > 0$ there is a $K_1 \in \mathcal{K}(M)$ with $K_1 \subseteq K$, $\mu(K \setminus K_1) < \varepsilon$ and $x|_{K_1}$ continuous. We recall the following basic results on measurable functions (see BOURBAKI [2], § IV.5 for a detailed study of Lusin measurability).

3.1. Proposition: 1) The pointwise limit of a sequence of measurable functions is measurable;
2) if E is separable, then $x : M \longrightarrow E$ is measurable provided that it is Borel-measurable (i.e. for each Borel subset A of E, $x^{-1}(A)$ is measurable).

We denote by $\mathcal{L}^\infty(M;F)$ the space of measurable functions x from M into F so that $M_\infty(x) < \infty$ where

$$M_\infty(x) := \inf\{ K > 0 : \|x(t)\| \leq K \text{ locally } \mu\text{-almost everywhere}\}.$$

M_∞ is a seminorm on $\mathcal{L}^\infty(M;F)$ and the corresponding quotient space is denoted by $(L^\infty(M;F), \| \|_\infty)$ - it is a Banach space. If $x \in L^\infty(M)$, $a \in F$ then $x \otimes a : t \longmapsto x(t)a$ is in $L^\infty(M;F)$ and $\|x \otimes a\|_\infty = \|x\|_\infty \|a\|$.

Now suppose that $(F, \| \|, \tau)$ is a complete Saks space. Once again, for simplicity, we shall always assume that τ is metrisable. Then F has a representation as a projective limit of a spectrum

$$\{\pi_n : F_{n+1} \longrightarrow F_n ; n \in \mathbb{N}\}$$

of Banach spaces. A function $x : M \longrightarrow F$ is <u>measurable</u> if for each $K \in \mathcal{K}(M)$, $\varepsilon > 0$, $n \in \mathbb{N}$, there is a $K \in \mathcal{K}_1(M)$ with $K_1 \subset K$, $\mu(K \setminus K_1) < \varepsilon$ and $\pi_n \circ x|_{K_1}$ continuous. We denote by $\mathcal{L}^\infty(M;F)$ the space of measurable functions x from M into F for which $M_\infty(x) < \infty$ where M_∞ is defined as above, using the norm $\| \ \|$. Once again, $L^\infty(M;F)$ denotes the associated normed space. We consider the following structures on $L^\infty(M;F)$:

$\| \ \|_\infty$ - the above norm;

τ^1_{loc} - the locally convex topology defined by the seminorms $\| \ \|^n_K : x \longmapsto \int_K \|\pi_n \circ x(t)\|_n \, d\mu$ as K runs through $\mathcal{K}(M)$ and n through $\mathbb{N}$.

Now if M is compact and F is a Banach space then $(L^\infty(M;F), \| \ \|_\infty, \| \ \|_K)$ is a complete Saks space (this follows once again from BOURBAKI [2], IV.3.4, Théorème 3 and 3.3.2) (Egoroff's theorem)). Now returning to the general situation, $\{L^\infty(K;F_n) : n \in \mathbb{N}, K \in \mathcal{K}(M)\}$ forms a projective system of Saks spaces and its projective limit is naturally identifiable with the Saks space $(L^\infty(M;F), \| \ \|_\infty, \tau^1_{loc})$. For any function in the latter space clearly defines a thread in the natural way. On the other hand, suppose that $\{x_{n,K} : n \in \mathbb{N}, K \in \mathcal{K}\}$ is a thread. First hold n fixed. By the localisation principle, the thread $\{x_{n,K} : K \in \mathcal{K}(M)\}$ defines a measurable function from M into F (BOURBAKI [2], IV.5.2, Proposition 2). Now as n varies, the resulting functions can be patched together to form a measurable function from M into F. The Saks space structure on $L^\infty(M;F)$ was so defined that this vector space isomorphism be a Saks space isomorphism. Hence we have:

3.2. <u>Proposition</u>: $(L^\infty(M;F), \| \|_\infty, \tau^1_{loc})$ is a complete Saks space.

The corresponding mixed topology is denoted by β_1. Then $(L^\infty(M;F), \beta_1)$ is a complete locally convex space.

3.3. <u>Lemma</u>: Suppose that M is compact and F is a Banach space. Then the unit ball of C(M;F) is τ^1_{loc}-dense in the unit ball of $L^\infty(M;F)$.

<u>Proof</u>: If $\varepsilon > 0$ and x is in the unit ball of $L^\infty(M;F)$, we can choose K_1 compact in M so that $\mu(M \setminus K_1) \le \varepsilon$ and $x|_{K_1}$ is continuous. Now we can extend $x|_{K_1}$ to a continuous function $\tilde{x}$ from M into F without increasing its supremum norm (generalised Tietze theorem). Then $\tilde{x}$ is in the unit ball of C(M;F) and $\|x - \tilde{x}\|_1 \le \varepsilon$.

3.4. <u>Proposition</u>: Let M be compact, F a Banach space. Then there is a natural Saks space isomorphism from $(L^\infty(M) \hat{\otimes}_\gamma F, \| \|, \tau^1_{loc} \hat{\otimes} \tau)$ onto $(L^\infty(M;F), \| \|, \tau^1_{loc})$.

<u>Proof</u>: There is a natural linear injection j from the algebraic tensor product $L^\infty(M) \otimes F$ into $L^\infty(M;F)$ and j is a contraction for both norms.

Now if $\sum_{i=1}^{n} a_i \chi_{M_i}$ is a step-function in $L^\infty(M;F)$ where the M_i's are disjoint, then

$$\|\Sigma\, a_i \chi_{M_i}\|_1 = \Sigma\|a_i\|\,\mu(M_i) \geq \|a_i \otimes \chi_{M_i}\|$$

and so j is an isometry for the L^1-norms.

If $\Sigma\, x_i \otimes a_i$ is a continuous function in $L^\infty(M) \otimes F$ then

$$\begin{aligned} \|\Sigma\, x_i \otimes a_i\|_\otimes &= \sup\{\|\Sigma\, f(a_i)x_i\| : f \in F', \|f\| \leq 1\} \\ &= \sup_{t \in M} \sup_{f \in F', \|f\| \leq 1} \{|\Sigma\, f(a_i)x_i(t)|\} = \\ &= \sup_{t \in M} \|\Sigma\, x_i \otimes a_i(t)\| = \|\Sigma\, x_i \otimes a_i\|_\infty \end{aligned}$$

and so j is an isometry on $C(M) \otimes F$ for the supremum norms.

Now the unit ball of $C(M) \otimes F$ is dense in the unit ball of $C(M;F)$ for the supremum norm (see SEMADENI [26], 20.5.6) and hence is dense in the unit ball of $L^\infty(M;F)$ in the L^1-norm (3.3). The result follows now from the construction of the completion of a Saks space (I.3.6).

3.5. Proposition: Let $(M;\mu)$ be a measure space, F a complete Saks space. Then there is a natural Saks space isomorphism from $(L^\infty(M) \hat{\otimes}_\gamma F, \|\;\|, \tau^1_{loc} \hat{\otimes} \tau)$ onto $(L^\infty(M;F), \|\;\|, \tau^1_{loc})$.

Proof: Consider the commutative diagram

$$\begin{array}{ccccc} L^\infty(M)\hat{\otimes}_\gamma F & \longrightarrow & L^\infty(K_1)\hat{\otimes}_\gamma F_m & \longrightarrow & L^\infty(K)\hat{\otimes}_\gamma F_n \\ \downarrow & & \downarrow & & \downarrow \\ L^\infty(M;F) & \longrightarrow & L^\infty(K_1;F_m) & \longrightarrow & L^\infty(K;F_n) \end{array}$$

$(K \subseteq K_1,\ n \leq m)$ where the vertical arrows are provided by 3.4.

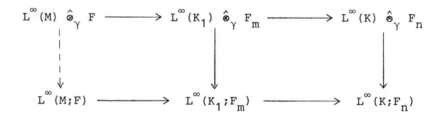

The required isomorphism is the induced arrow from $L^{\infty}(M) \hat{\otimes}_{\gamma} F$ into $L^{\infty}(M;F)$.

We now consider duality theory for $L^{\infty}(M;F)$. To avoid technicalities, we shall simplify even further and suppose for the rest of this section that M is compact and E is a Banach space. In fact, the main analytic difficulties occur already in this situation. A first guess might be that the β_1-dual of $L^{\infty}(M;F)$ is $L^1(M;F')$. It turns out that this is not true in general and that the true story is rather complicated. In fact, the question of when the dual of $(L^{\infty}(M;F),\beta_1)$ is $L^1(M;F')$ is closely related to the Radon-Nikodym property.

In order to apply I.1.18 we must first describe the dual space of $L^1(M;F)$.

3.6. <u>Proposition</u>: For any $x \in L^1(M;F)$, $f \in L^{\infty}(M;F')$, the scalar function $\langle x,f \rangle$ is in $L^{\infty}(M;\mu)$ and the bilinear form

$$(x,f) \longmapsto \int_M \langle x,f \rangle \, d\mu$$

induces an isometry from $L^{\infty}(M;F')$ into $L^1(M;F)'$.

<u>Proof</u>: If K_1 and K_2 are compact so that $\mu(M \setminus K_1) < \varepsilon/2$. $\mu(M \setminus K_2) < \varepsilon/2$ and x (resp. f) is continuous on K_1 (resp. K_2) then $\langle x,f \rangle$ is continuous on $K_1 \cap K_2$ and $\mu(M \setminus (K_1 \cap K_2)) < \varepsilon$. Hence this function is measurable.

We also have the estimate

$$\int_M |\langle x,f\rangle|\,d\mu \le \|f\|_\infty \int_M \|x\|\,d\mu = \|f\|_\infty \|x\|_1$$

Hence f defines a functional T_f in $L^1(M;F)'$ and the above inequality shows that

$$\|T_f\| \le \|f\|_\infty .$$

To prove the reverse inequality, we assume first that f is countably valued i.e. that f has the form $\Sigma f_n \otimes \chi_{A_n}$ where (f_n) is a sequence in F' and (A_n) is a disjoint sequence of measurable subsets (of non-zero measure).

If $\varepsilon > 0$ is given we choose n_0 such that $\|f_{n_0}\| + \varepsilon/2 \ge \|f\|_\infty$ (for $\|f\|_\infty = \sup_n \|f_n\|$).
Now let $x_{n_0} \in F$ be such that $\|x_{n_0}\| = 1$ and $|f_{n_0}(x_{n_0})| \ge \|f_{n_0}\| - \varepsilon/2$. Then $x_{n_0} \otimes \chi_{A_{n_0}}$ is an element of $L^1(M;E)$ of norm $\mu(A_{n_0})$ and

$$\begin{aligned} T_f(x_{n_0} \otimes \chi_{A_n}) &= \int_M \langle x_{n_o} \otimes \chi_{A_{n_o}}, f\rangle\,d\mu = f_{n_0}(x_0)\mu(A_{n_0}) \\ &\ge (\|f\|_\infty - \varepsilon)\,\|x_{n_0} \otimes \chi_{A_{n_o}}\| \end{aligned}$$

This shows that $\|T_f\| \ge \|f\|_\infty$ for countably valued f.
Now the countably valued f are norm dense in $L^\infty(M;F')$ and from this it follows easily that the above inequality is valid for all f in $L^\infty(M;F')$.

The interesting question is, of course, when this mapping is surjective (i.e. $(L^1(M;F)' = L^\infty(M;F'))$. Now this occurs exactly when F' has the Radon-Nikodym property (discussed in 3.15 below) and so the above question leads to problems outside the scope of this book. One can, however, give a fairly elementary proof in the special case where F has a separable dual. We begin with some Lemmas:

3.7. <u>Lemma</u>: Let E be a separable Banach space, F a countable, norm-determining subset of E'. Then the σ-algebra A generated by F on E is precisely the Borel algebra of E (i.e. the σ-algebra generated by the norm open sets).

<u>Proof</u>: We recall that F is norm-determining if, for each $x \in E$, $\|x\| = \sup \{ |f(x)| : f \in F, \|f\| \le 1\}$.
Then the closed unit ball of E is

$$\{x \in E : |f(x)| \le 1 \quad \text{for } f \in F, \|f\| \ge 1\}$$

and so is in A. Therefore the same holds for the open unit ball (since it is a countable union of closed unit balls). Now every open set in E is a countable union of open balls and so is in A.

3.8. <u>Corollary</u>: Let E and F be as above and let Φ be a function from M into E so that for each $f \in F$, $f \circ \Phi$ is measurable. Then Φ is Lusin measurable.

<u>Proof</u>: It follows immediately from 3.7 that Φ is Borel measurable. The result follows then from 3.1.2).

3.9. Definition: If E is a subspace of $L^{\infty}(M;\mu)$, then a lifting for E is a linear mapping $\rho : E \longrightarrow \mathcal{L}^{\infty}(M;\mu)$ so that

1) $\pi \circ \rho = Id_E$ where π is the natural projection from $\mathcal{L}^{\infty}$ onto L^{∞};

2) $|\rho(x)(t)| \leq \|x\|_{\infty}$ for $t \in M$.

3.10. Proposition: If E is a norm separable subspace of $L^{\infty}(M;\mu)$, then E has a lifting.

Proof: E contains a countable dense subset E_0 which is a vector space over the rationals $\mathbb{Q}$. Let (x_n) be a basis for E_0 (over $\mathbb{Q}$). For each $n \in \mathbb{N}$ let $\tilde{x}_n$ be an element of $\mathcal{L}^{\infty}$ so that $\pi(\tilde{x}_n) = x_n$. Then we construct a $\mathbb{Q}$-linear mapping $\tilde{\rho}$ from E_0 into $\mathcal{L}^{\infty}$ which maps x_n into $\tilde{x}_n$. For each $x \in E_0$, $|\tilde{\rho}(x)(t)| \leq \|x\|_{\infty}$ except for a locally negligable set A_x. Then for each $x \in E_0$ we let $\tilde{\tilde{x}}$ be the function which is equal to $\tilde{\rho}(x)$ for $t \notin \bigcup_{x \in E_0} A_x$ and which is zero on the latter set. Then the mapping $\rho : x \longrightarrow \tilde{\tilde{x}}$ is $\mathbb{Q}$-linear and satisfies the condition

$$|\rho(x)(t)| \leq \|x\|_{\infty} \text{ for } x \in E_0,\ t \in M.$$

We can extend ρ by continuation to obtain the required lifting.

3.11. Proposition: Let E be a Banach space so that E' is separabl[e]. Then the mapping constructed in 3.6 is surjective (i.e. the dual of $L^1(M;E)$ is $L^{\infty}(M;E')$).

Proof: Let ϕ be an element of $L(M;E)'$. Then for each $x \in E$, the mapping

$$T_\phi(x) : y \longrightarrow \phi(y \otimes x)$$

is an element of the dual of $L^1(M;\mu)$ and so is represented by a function in $L^\infty(M;\mu)$. Hence we can regard T_ϕ as a mapping from E into $L^\infty(M;\mu)$. It is clearly linear and has norm $\leq \|\phi\|$. Now the range of T_ϕ is a separable subspace of $L^\infty(M;\mu)$ (since E is separable) and so there is a lifting ρ on it.

Now if $t \in M$, we define $f(t)$ to be the composition

$$E \xrightarrow{T_\phi} L^\infty \xrightarrow{\rho} \mathcal{L}^\infty \longrightarrow \mathbb{C}$$

where the last arrow is evaluation at t. Then $f(t) \in E'$ and its norm is $\leq \|\phi\|$. Hence we have constructed a bounded function $f : M \longrightarrow E'$ and for any $x \in E$, $\langle x,f \rangle$ is measurable. Hence we can apply 3.8 to deduce that f is Lusin measurable (note that any countable dense subset of E is norm-determining as a subset of the dual of E').

To complete the proof, we must show that for any $x \in L^1(M;E)$ we have

$$\phi(x) = \int \langle x,f \rangle d\mu.$$

It suffices to show this for the case where x has the form $y \otimes z$ ($y \in L^1(M;\mu)$, $z \in E$). Then we can calculate:

$$\phi(y \otimes z) = T_\phi(z)(y) = \int y\, T_\phi(z) d\mu$$

$$= \int y \langle z,f \rangle d\mu = \int \langle y \otimes z, f \rangle d\mu$$

Using this result, we can obtain immediately from I.1.18 the following result:

3.12. Proposition: Let F be a Banach space whose dual is separable. Then the natural duality bewteen $L^\infty(M;F)$ and $L^1(M;F')$ establishes an isometry from $L^1(M;F')$ onto the β_1-dual of $L^\infty(M;F)$.

Before stating the next result, we remark that the definition of equi-integrability (given before 1.2) can be extended to functions with values in a Banach space simply by replacing absolute values by norms.

3.13. Proposition: When F is as in 3.12 and H is a subset of $L^1(M;E')$, then the following are equivalent:

1) H is β_1-equicontinuous;
2) H is relatively $\sigma(L^1(M;F'), L^\infty(M;F))$-compact;
3) H is norm-bounded and 1-equi-integrable;

Proof: 1) $\Longrightarrow$ 2) and 3) $\Longrightarrow$ 1) can be proved exactly as in the scalar case.

2) $\Longrightarrow$ 3): H is clearly norm-bounded. If H were not 1-equi-integrable, then by 1.7 (applied to the set $\{\|f\| : f \in H\}$ in $L^1(M;\mu)$ there would be a disjoint sequence (A_n) of measurable sets, a sequence (f_n) in H and an $\varepsilon > 0$ so that

$$\int_{A_n} \|f_n\| \, d\mu \geq \varepsilon \qquad (n \in \mathbb{N}).$$

For every $n \in \mathbb{N}$ choose $y_n \in L^\infty(M;F)$ such that y_n is supported by A_n, $\|y_n\|_\infty \leq 1$ and

$$f_n(y_n) = |\int_{A_n} \langle f_n, y_n \rangle d\mu| \geq \varepsilon/2$$

(this can be done by approximating f_n by simple functions). Now consider the mapping

$$T : f \longmapsto (f(y_n))_{n=1}^\infty$$

from $L^1(M;F')$ into $\ell^1(\mathbb{N})$. Then T is continuous but $\{T f_n\}$ is not weakly compact - contradiction.

3.14. Corollary: Under the assumption of 3.12, $(L^\infty(M;E), \beta_1)$ is a Mackey space.

3.15. Remarks on the Radon Nikodym property: If we compare 2.4 with the scalar result, we are led naturally to pose the following question: when can we identify the space $M_{bv}^\nu(\Sigma;E)$ of ν-continuous measures of bounded variation with the space $L^1(M;E)$ i.e. when does every E-valued ν-continuous measure of bounded variation have a ν-derivative? The following example shows that this is not always the case:

Example: Our measure space is [0,1] with Lebesgue measure. We define a vector-valued measure μ by defining $\mu(A)$ to be the sequence $(\int_A \sin(2\pi nt)d\nu(t))_{n=1}^\infty$. Then μ takes its values in c_o (RIEMANN-LEBESGUE Lemma) and it is routine to show that it is σ-additive and ν-continuous. Now if μ has a ν-derivate, then it must be the function $t \longmapsto (\sin(2\pi nt))$. However, this function does not take values in c_o.

We formalise the above fact in the following definition:

Definition: A Banach space E has the Radon Nikodym property (abbreviated RNP) if for each measure space $(M;\nu)$ the canonical injection

$$L^1(M;E) \longrightarrow M_{bv}^{\nu}(\Sigma;E)$$

is surjective. We remark that it is sufficient to check this condition for the canonical measure space [0,1]. The above example shows that c_o does not have the RNP. As a positive result we have:

Let E be a Banach space with a boundedly complete basis (x_n). Then E has the RNP.

We sketch the proof (as we shall see later, this result follows from 3.12). It is convenient to renorm E so that

$$\|\sum_{k=1}^{\infty} \lambda_k x_k\| = \sup_n \|\sum_{k=1}^{n} \lambda_k x_k\|.$$

Now let μ be an E-valued measure which is ν-continuous. Then the induced measures $\{f_n \circ \mu\}$ (where (f_n) is the associated biorthogonal sequence) are ν-continuous and so there are measurable functions y_n so that

$$f_n \circ \mu(A) = \int_A y_n d\mu \qquad (A \in \Sigma).$$

Now it follows from the fact that the basis is boundedly complete that the E valued series $\Sigma\, y_n x_n$ converges μ-almost everywhere to a measurable function (since the simple estimate

$\int_A \|\sum_{k=1}^{n} y_k(t)x_k\| \, d\mu(t) \le |\mu|(A)$ shows that the partial sums are bounded almost everywhere). $|\mu|$ is the variation of μ (i.e. $|\mu|(A) = \sup \Sigma \|F(A_n)\|$; the supremum being taken over the finite partitions of A). The limit is a ν-derivative for μ.

For a detailed study of the Radon-Nikodym property, we refer to DIESTEL and UHL [8] and BUCHWALTER and BUCCHIONI [3].

For the reader's orientation, we quote the following results:

I. Any reflexive space and any separable (or even weakly compactly generated) dual space has the RNP. In fact, the dual of a separable Banach space has RNP iff it is separable.

II. A Banach space has the RNP if and only if every separable subspace has the RNP.

III. The spaces $L^1(M;\mu)$, $C(K)$ and $L^\infty(M;\mu)$ do not have the RNP (except for the trivial exceptions - μ atomic, K finite and μ of finite support resp.).

In fact, Proposition 3.12 is a special case of the following result: the space $L^1(M;E')$ is naturally isometric to the β_1 dual of $L^\infty(M;E)$ if and only if E' has the Radon-Nikodym property.

Hence we can interpret 3.12 as stating that a separable dual space has the RNP (and this result contains the result that a space with a boundedly complete basis has the RNP of course). In the general case, one can identify the dual of $(L^\infty(M;E),\beta_1)$ with the lower star space $L^1_*(M;E')$ introduced by SCHWARTZ (roughly speaking, this is the space of those functions x which are weak-star measurable and are such that the upper integral of the norm function $\|x\|$ is finite). Since a reflexive space has the RNP, 3.12 and its Corollaries hold for such spaces.

III.4. NOTES

The topology β_σ has never been studied before (of course, 1.1.1) identifies it with a familiar topology of the dual pair (L^∞, L^1)). The study of β_1 was suggested by GROTHENDIECK's paper [12] where the version of the DUNFORD-PETTIS theorem proved here (1.8) can be found. 1.4 is due to GODEMENT. Many of the results of § 1, in particular 1.9, were obtained independently by DAZORD and JOURLIN [7]. The result that (L^∞, β_1) is a Mackey space has also been obtained by STROYAN (for bounded measures). The remark that (ℓ^∞, β_1) is a universal Schwartz space is due to JARCHOW [15]. SENTILLES [27] has also introduced a strict topology for L^∞-spaces in his study of a measure free approach to L^1-spaces based on the Boolean algebra of measurable sets. For a complete account of the integration theory used in 1.20 and in particular, for the definition of $\tilde{L}^\infty(M;\Sigma)$ see the lecture notes of BUCHWALTER and BUCCHIONI [3].

Our basic reference for § 2 was DIESTEL and UHL ([8], Ch. 1) - in particular, 2.2 and 2.3 can be found there.

As mentioned in the introduction, § 3 is devoted to the elementary beginnings of a theory of measurable functions with values in a

Saks space. We refer the reader to SCHWARTZ [25] for a discussion of the problems involved in the duality theory for Banach space valued L^p-functions.

The final form of this chapter owes a great deal to the assistance of W. SCHACHERMAYER. In addition to simplifying and improving some proofs he pointed out the important references [3] and [8] to me. Paragraphs 1.20. 3.6 - 3.13 are based on [23] and [24].

REFERENCES FOR CHAPTER III.

[1] T. ANDO — Weakly compact sets in Orlicz spaces, Can. J. Math. 14 (1962) 170-176.

[2] N. BOURBAKI — Eléments de Mathématique: Intégration, Chs. 1-4 (Paris, 1965).

[3] H. BUCHWALTER, D. BUCCHIONI — Intégration vectorielle et théorème de Radon-Nikodym (Département de Mathematiques, Lyon, 1975).

[4] A.K. CHILANA — The space of bounded sequences with mixed topology, Pac. J. Math. 48 (1973) 29-33.

[5] H.S. COLLINS — On the space $\ell^\infty(S)$, with the strict topology, Math. Z. 106 (1968) 361-373.

[6] J.B. CONWAY — Subspaces of $(C(S),\beta)$, the space (ℓ^∞,β) and (H^∞,β) Bull. Amer. Math. Soc. 72 (196 79-81.

[7] J. DAZORD, M. JOURLIN — Une topologie mixte sur l'espace L^∞, Publ. Dép. Math. Lyon 11-2 (1974) 1-18.

[8] J. DIESTEL, J.J. UHL, Jr. — The theory of vector measures (to appear in the American Mathematical Society's surveys).

[9] J. DIXMIER — Les algèbres d'opérateurs dans l'espace hilbertien (Paris, 1969).

[10] D.H. FREMLIN — Pointwise compact sets of measurable functions, Man. Math. 15 (1975) 219-242.

[11] A. GOLDMAN — L'espace des fonctions localement L^p sur un espace compactologique ✶ (Preprint).

[12] A. GROTHENDIECK — Sur les applications linéaires faiblement compactes d'espaces du type C(K), Can. J. Math. 5 (1953) 129-173.

[13] E.A. HEARD Kahane's construction and the weak sequential completeness of L^1, Proc. Amer. Math. Soc. 44 (1974) 96-100.

[14] A. IONESCU TULCEA On pointwise convergence and equicontinuity in the lifting topology I,II (Z. Wahrsch. u. verw. Gebiete 26 (1973) 197-205 and Adv. in Math. 12 (1974) 171-177.

[15] H. JARCHOW Die Universalität des Raumes c_0. Math. Ann. 203 (1973) 211-214.

[16] J.W. JENKINS On the characterization of abelian W*-algebras, Proc. Amer. Math. Soc. 35 (1972) 436-438.

[17] S.S. KHURANA A vector form of Phillip's Lemma, J. Math. Anal. Appl. 48 (1974) 666-668.

[18] Weakly convergent sequences in L^∞ (Preprint).

[19] V.L. LEVIN Lebesgue decomposition for functionals on the space L^∞_* of vector-valued functions, Func. Appl. 8 (1974) 314-317.

[20] D.R. LEWIS Conditional weak compactness in certain inductive tensor products, Math. Ann. 201 (1973) 201-209.

[21] S. SAKAI C*-algebras and W*-algebras (Berlin, 1971).

[22] S. SAKS On some functionals, Trans. Amer. Math. Soc. 35 (1933) 549-556.

[23] W. SCHACHERMAYER Mixed topologies on L^∞ for abstract measures (written communication).

[24] Mixed topologies in spaces of Banach valued functions (written communication).

[25] L. SCHWARTZ Fonctions mésurables et *-scalairement mésurables, mesures banachiques majorées, martingales banachiques et propriété de Radon Nikodym, Sém. Maurey-Schwartz 1974/75, Exp. 4.

[26] Z. SEMADENI Banach spaces of continuous functions I, (Warsaw, 1971).

[27] F.D. SENTILLES An L^1-space for Boolean algebras and semi-reflexivity of spaces $L^\infty(M,\Sigma,\mu)$ (Preprint).

[28] K.D. STROYAN A characterisation of the Mackey uniformit $m(L^\infty,L^1)$ for finite measures, Pac. J. Math 49 (1973) 223-228.

[29] E. THOMAS The Lebesgue-Nikodym Theorem for vector valued Radon measures, Memoirs of the American Math. Soc. 139 (1974).

[30] J. VESTERSTRØM, W. WILS On point realization of L^∞-endomorphisms, Math. Scand. 25 (1969) 178-180.

[31] C.R. WARNER Weak * dense subspaces of $L^\infty(\mathbb{R})$, Math. Ann. 197 (1972) 180-181.

CHAPTER IV - VON NEUMANN ALGEBRAS

Introduction: In this chapter, we study algebras of operators on Hilbert space from the point of view of mixed topologies. On the algebra L(H) (H a Hilbert space) there are three classical topologies - corresponding to uniform, strong and weak convergence of operators. For most applications (in particular, to spectral theory), the uniform topology is too strong. From a theoretical point of view, the weak and strong topologies are very unsatisfactory. VON NEUMANN introduced rather complicated topologies - the ultraweak and ultrastrong topologies - to overcome these difficulties. We introduce here three natural mixed topologies on L(H) which are related to (but distinct from) the ultraweak and ultrastrong topologies and have several advantages with respect to these topologies.

There are two distinct approaches to algebras of operators: the direct approach by defining them as self-adjoint, ultraweakly closed subalgebras of L(H) (von Neumann algebras) or the axiomatic approach (W^*-algebras). Comprehensive treatments can be found in the books of DIXMIER [8] and SAKAI [14] respectively. The latter approach is certainly the most elegant. However we have chosen the former since it allows us to give a more elementary treatment. In fact, with the exception of one rather deep theorem of SAKAI, we require only acquaintance with spectral theory of self-adjoint operators and (what is essentially the same) the representation theory of commutative von Neumann algebras.

In § 1 we consider two mixed structures on L(H) and identify its dual as the space of nuclear operators on H. We then present the corresponding Saks spaces as a projective limit of spaces of finite dimensional operators and deduce that L(H) has the approximation property.

In the second section, we consider three mixed topologies β_σ, β_s and β_{s*} on a von Neumann algebra. For commutative von Neumann algebras (i.e. L^∞-spaces) these coincide with the topologies considered in Ch. III (in this case, $\beta_s = \beta_{s*} = \beta_1$). After giving routine properties of these topologies, we use them to give a short proof of KAPLANSKY's density theorem. The main result is that β_{s*} is a Mackey topology and this has several important consequences. The section ends with some brief remarks on the relation between the mixed topologies and classical topologies on von Neumann and W*-algebras.

IV.1. THE ALGEBRA OF OPERATORS IN HILBERT SPACES

Let L(H) denote the algebra of bounded, linear operators on a Hilbert space H. On L(H) we consider the following structures:

- $\| \|$ - the uniform norm;
- τ_s - the topology of pointwise convergence on H with respect to the norm (<u>the strong operator topology</u>);
- τ_σ - the topology of pointwise convergence on H with respect to the weak topology $\sigma(H,H')$ (<u>the weak operator topology</u>).

Note that we can consider L(H) as a vector subspace of $C^\infty(B_{\| \|};H)$ and it is a locally convex subspace for the following pairs of structures:

$$(L(H),\| \|) \; : \; (C^\infty(B_{\| \|};H),\| \|_\infty)$$
$$(L(H),\tau_s) \; : \; (C^\infty(B_{\| \|};H),\tau_p)$$
$$(L(H),\tau_\sigma) \; : \; (C^\infty(B_{\| \|};H_\sigma),\tau_p)$$

where H_σ denotes the Saks space $(H,\| \|, \sigma(H,H'))$.

1.1. <u>Proposition</u>: The unit ball of L(H) is complete for τ_s and compact for τ_σ.

Hence we have two complete Saks spaces available:

$$(L(H),\| \|,\tau_s) \quad \text{and} \quad (L(H),\| \|,\tau_\sigma).$$

We denote the corresponding mixed topologies by β_s and β_σ. Note that the above spaces are Saks space subspaces of $(C^\infty(B_{\|\ \|};H),\|\ \|_\infty,\tau_K)$ and $(C^\infty(B_{\|\ \|};H_\sigma),\|\ \|_\infty,\tau_K)$ and that $(L(H),\beta_\sigma)$ is a locally convex subspace of $(C^\infty(B_{\|\ \|};H),\beta_K)$ (I.4.4 and 1.1).

We shall consider the properties of the above locally convex spaces in more detail in the next section. In the following we consider duality for L(H). We require two preliminary results:

1.2. Lemma: Let E and F be locally convex spaces and let $(L(E;F),\tau_p)$ be the space of continuous linear operators from E into F with the topology of pointwise convergence on E. Then the dual of $(L(E;F),\tau_p)$ is identifiable with $E \otimes F'$ under the bilinear form

$$(T, \sum_{i=1}^{n} f_i \otimes x_i) \longmapsto \sum_{i=1}^{n} f_i(Tx_i).$$

Proof: We regard $(L(E;F),\tau_p)$ as a topological subspace of $(C(E;F),\tau_p)$. Then if $f \in (L(E;F),\tau_s)'$ there is an extension of f to an element of the dual of C(E;F). Such an element has the form

$$T \longmapsto \sum_{i=1}^{n} f_i(Tx_i)$$

for some finite subsets $\{x_1,\dots,x_n\}$ of E and $\{f_1,\dots,f_n\}$ of F'.

1.3. Lemma: An operator $A \in L(H)$ is compact if and only if there exist orthogonal sequences (x_n) in H, (f_n) in H' and a positive element (λ_n) of $c_0(\mathbb{N})$ so that

$$A : x \longmapsto \sum_n \lambda_n f_n(x) x_n$$

(i.e. $A = \sum_n \lambda_n f_n \otimes x_n$).

Proof: We shall prove that every compact operator has this form. The operator $A^*A : H \longrightarrow H$ is positive and compact. Hence, by the spectral theorem for compact operators, there is an orthogonal sequence (x_n) in H and a positive sequence (λ_n^2) in $c_0(\mathbb{N})$ so that

$$A^*A : x \longmapsto \sum_n \lambda_n^2 (x|x_n) x_n.$$

Let M be the orthogonal complement of $A^*A(H)$. Then $A|_M = 0$ since if $y \in M$

$$||Ay||^2 = (Ay|Ay) = (y|A^*Ay) = 0.$$

Then a simple calculation shows that the sequence (f_n) in H' is orthonormal (where $f_n : x \longmapsto (\lambda_n^{-1} x|Ax_n))$ and that $A = \Sigma \lambda_n x_n \otimes f_n$.

1.4. Definition: The above proof shows that the positive sequence (λ_n) is an invariant of A (i.e. independent of the particular representation of A - up to permutations of course), since it is the sequence of the square roots of the eigenvalues of A^*A. Hence we can make the following definition:

A compact operator $A : H \longrightarrow H$ is <u>nuclear</u> if $\Sigma \lambda_n < \infty$. We denote the latter sum then by $\|A\|_N$.

1.5. <u>Proposition</u>: If $A = \Sigma \lambda_n f_n \otimes x_n$ is a nuclear operator in L(H) then

$$f_A : T \longmapsto \sum_n \lambda_n f_n(Tx_n)$$

is a continuous linear form on $(L(H), \| \ \|)$ and $\|f_A\| = \|A\|_N$.

<u>Proof</u>: We verify the statement about the norms. Firstly we can estimate

$$|f_A(T)| \leq \sum_n \lambda_n \|f_n\| \|T\| \|x_n\| = (\sum_n \lambda_n)(\|T\|)$$

and so $\|f_A\| \leq \|A\|_N$.

On the other hand, there is a partial isometry V on H so that $Vx_n = y_n$ for each n where (y_n) is a sequence in H, biorthogonal to (f_n). Then $\|V\| \leq 1$ and $f_A(V) = \Sigma \lambda_n$.

Now it is easily checked that the mapping $A \longmapsto f_A$ is well-defined (i.e. the linear form f_A is independent of the representation of A). For if $\Sigma \lambda_n f_n \otimes x_n = 0$ then for each $m \in \mathbb{N}$,

$$\lambda_m x_m = (\sum_n \lambda_n f_n \otimes x_n)(y_m) = 0$$

where (y_n) is biorthogonal to (f_n) and so each $\lambda_m = 0$. Hence we have embedded N(H), the space of nuclear mappings in L(H), isometrically in $(L(H), \| \ \|)'$.

1.6. <u>Proposition</u>: $(N(H), \| \|_N)$ is the dual of the spaces $(L(H), \beta_s)$ and $(L(H), \beta_\sigma)$.

<u>Proof</u>: We can identify $H \otimes H'$, the dual of $(L(H), \tau_s)$ and $(L(H), \tau_\sigma)$ (1.3), with the space of operators in $L(H)$ of finite rank. Now these are obviously $\| \|_N$-dense in $N(H)$ (for the operators $\sum_{n=1}^{m} \lambda_n f_n \otimes x_n$ approximate $\sum_{n=1}^{\infty} \lambda_n f_n \otimes x_n$ in the nuclear norm) and so the result follows from I.1.18.(ii).

1.7. <u>Remark</u>: If $A \in L(H)$ we denote by $Tr(A)$ the <u>trace</u> of A i.e. the sum $\sum_{\alpha \in A} (Ax_\alpha | x_\alpha)$ where $(x_\alpha)_{\alpha \in A}$ is an orthonormal basis for H if this sum is absolutely convergent. Then $Tr(A)$ is independent of the basis (x_α) and $N(H)$ is precisely the class of operators in $L(H)$ for which the trace exists (hence the nuclear operators are sometimes called <u>operators of trace class</u>). One can then verify that if $A = \sum \lambda_n f_n \otimes x_n \in N(H)$ and $T \in L(H)$ then $Tr(AT)$ exists and is given by

$$Tr(AT) = \sum_n \lambda_n f_n(Tx_n) = f_A(T)$$

and so the duality between $N(H)$ and $L(H)$ is given by the bilinear form

$$(A,T) \longmapsto Tr(AT).$$

We now give a natural projective limit representation of the Saks space $(L(H), \| \|, \tau_\sigma)$ and $(L(H), \| \|, \tau_s)$. We order the family of pairs (M,N) of closed subspaces of H by defining

$$(M,N) \leq (M_1,N_1) \Longleftrightarrow M \subseteq M_1 \quad \text{and} \quad N \subseteq N_1.$$

If $(M,N) \leq (M_1,N_1)$ there is a natural contraction from $L(M_1,N_1)$ into $L(M,N)$ defined by

$$T \longmapsto \pi \circ T \circ i$$

where $i : M \longrightarrow M_1$ is the natural injection and $\pi : N_1 \longrightarrow N$ the orthogonal projection. Then if F denotes the family of all finite dimensional subspaces of H,

$$\{L(M,N) : M,N \in F\}$$

$$\{L(M,H) : M \in F\}$$

form projective systems of Banach spaces under these mappings. Their Banach space projective limits are identifiable with L(H) and their Saks space projective limits are

$$(L(H),\| \|,\tau_\sigma) \quad \text{and} \quad (L(H),\| \|,\tau_s)$$

respectively.

Now the above mapping from $L(M_1,N_1)$ into $L(M,N)$ has a contractive right inverse, namely the mapping

$$T \longrightarrow i \circ T \circ \pi$$

Hence condition b) of Proposition I.4.20 is satisfied. If M and N are subspaces of H and M is finite dimensional then the Banach space L(M,N) has the metric approximation property. Hence we can apply I.4.20 to obtain the following result:

1.8. Proposition: $(L(H),\beta_\sigma)$ and $(L(H),\beta_s)$ have the approximation property.

We remark here that it has apparently been shown recently that $(L(H),\| \|)$ does not have the approximation property (SZANKOWSKI).

IV.2. VON NEUMANN ALGEBRAS

2.1. Proposition: Let E be a subspace of L(H). Then the following are equivalent:

1) E is β_s-closed;
2) E is β_σ-closed;
3) the unit ball of E is β_σ-compact.

Proof: The equivalence of 1) and 2) follows from the fact that β_s and β_σ are topologies of the same duality, that of 2) and 3) from I.4.3.

2.2. Definition: Let H be a Hilbert space. A von Neumann algebra on H is a $*$-subalgebra M of L(H) which satisfies one of the conditions of 2.1 and contains the unit Id_H of L(H). Then we can regard M as a Saks space in two natural ways: as a subspace of $(L(H),\| \|,\tau_s)$ resp. of $(L(H),\| \|,\tau_\sigma)$. We denote the corresponding mixed topologies by β_s and β_σ.

At this point it is convenient to introduce a symmetric form of β_s. We strengthen the topology τ_s by defining τ_{s^*} to be the topology defined by the seminorms $\{p_x, p_x^* : x \in H\}$ where

$$p_x : T \longmapsto \|Tx\| \; ; \qquad p_x^* : T \longmapsto \|T^*x\|$$

Then $(L(H), \| \|, \tau_{s^*})$ is a complete Saks space and the associated mixed topology is denoted by β_{s^*}. Note that the dual of $(L(H), \beta_{s^*})$ is also $N(H)$ (this follows from 1.6 and the fact that the involution is β_σ-continuous). The topology β_{s^*} has the following advantage over β_s : if we denote by M^s the set of self-adjoint elements of a von Neumann algebra M then M^s is a vector space over the reals and M can be identified (as a vector space) with the complexification of M^s. Now $\beta_s = \beta_{s^*}$ on M^s and the locally convex complexification of (M^s, β_s) is (M, β_{s^*}). Hence to verify for example the β_{s^*}-continuity of a linear mapping on M, it is sufficient to check its β_s-continuity on M^s.

For convenience we collect in one Proposition the basic properties of the above topologies:

2.3. <u>Proposition</u>: Let M be a von Neumann algebra. Then

1) (M, β_σ), (M, β_s) and (M, β_{s^*}) are complete;

2) the dual of M under any of the above topologies is the Banach space $M_* := N(H)/M^o$;

3) a linear operator Φ from M into a locally convex space is β_σ- (resp. β_s-, resp. β_{s*}-) continuous if and only if its restriction to the unit ball B_M of M is τ_σ- (resp. τ_s-, resp. τ_{s*}-) continuous;

4) if M is of countable type (i.e. if H contains a countable subset S so that for each $T \in M$, $T|_S = 0 \Longrightarrow T = 0$), then the unit ball of M is metrisable for the topologies β_σ, σ_s and β_{s*};

5) (M,β_σ) is semi-Montel and so is the dual of the Banach space M_*, with the topology of compact convergence. Hence a subset A of (M,β_σ) is closed if and only if $A \cap rB_M$ is τ_σ-closed for each $r > 0$;

6) if M is of countable type, a subset A of M is β_σ-closed if and only if it is sequentially closed;

7) multiplication is jointly β_{s*}-continuous on B_M;

8) β_s and τ_s (resp. β_{s*} and τ_{s*}) have the same convergent sequences and compact sets;

9) β_σ is defined by the seminorms

$$T \longmapsto \sup_{n \in \mathbb{N}} \lambda_n^{-1} |\mathrm{Tr}(A_n T)|$$

where (A_n) runs through the sequences in the unit ball of N(H) and (λ_n) runs through the family of sequences of positive numbers which increase to infinity;

10) if M_1 is a von Neumann algebra with $M_1 \subseteq M$ then β_σ induces β_σ on M_1.

If M is a von Neumann algebra, a linear operator from M into a locally convex space is normal if and only if it is β_σ-continuous (it is then β_{s^*}-continuous as we shall see below). Note that M_* is the space of normal mappings from M into $\mathbb{C}$. A $*$-algebra homomorphism from M into a von Neumann algebra M_1 is normal if it is β_σ-continuous (note that it is automatically a norm contraction - see DIXMIER [8], p. 8). Once again, it is equivalent to demand β_{s^*}-continuity. By I.4.32, it suffices that the homomorphism have a β_σ-closed graph. Two von Neumann algebras are isomorphic if there is a bijective normal $*$-algebra homomorphism from M onto M_1 (its inverse is then normal).

We recall now that if $(M;\mu)$ is a measure space then $L^\infty(M;\mu)$ can be regarded (via multiplication) as a subalgebra of $L(H)$ where H is the Hilbert space $L^2(M;\mu)$ and so as a (commutative) von Neumann algebra. Then the topology β_σ introduced here coincides with the β_σ of § III.1 and

$$\beta_s = \beta_{s^*} = \beta_1$$

on L^∞. The representation theorem for commutative von Neumann algebras states that every such algebra is isomorphic to an algebra of the above form.

The spectral theorem states that if $T \varepsilon L(H)$ is self-adjoint then there is a partition of unity $\{E(\lambda)\}$ on $\sigma(T)$ so that

$$T = \int_{\sigma(T)} \lambda \, dE(\lambda).$$

Then if $h \in L^{\infty}(\sigma(T))$ we can define

$$h(T) := \int_{\sigma(T)} h(\lambda)dE(\lambda).$$

The mapping $h \longmapsto h(T)$ is a normal $*$-algebra homomorphism from $L^{\infty}(\sigma(T))$ into $L(H)$. If T lies in a von Neumann subalgebra M of $L(H)$ then the range of $h \longmapsto h(T)$ lies in M (since $p(T) \in M$ for any polynomial p and the polynomials are β_1-dense in $L^{\infty}(\sigma(T))$). In particular, the projections $\{E(\lambda)\}$ in the spectral representation of T lie in M (since $E(\lambda) = \chi_{]-\infty,\lambda]\cap\sigma(T)}(T)$).

The following Proposition allows us to give a short proof of the KAPLANSKY density theorem:

2.4. <u>Proposition</u>: Let h be a bounded continuous function on $\mathbb{R}$. Then $T \longmapsto h(T)$ is β_s-continuous from B_{M^s} into $\|h\|_\infty B_{M^s}$.

<u>Proof</u>: It follows from 2.3.7) that the result holds for functions h which are polynomials on a neighbourhood of $[-1,1]$. Now take $T \in B_{M^s}$ and let $h(T) + U$ be a β_s-neighbourhood of $h(T)$. Then there is an $\varepsilon > 0$ so that $\varepsilon B_{M^s} \subseteq U$. Let $\tilde{h}$ be a function so that $\tilde{h}$ is a polynomial on a neighbourhood of $[-1,1]$ and $\|\tilde{h}\|_\infty \leq \|h\|_\infty$, $\|h - \tilde{h}\|_\infty \leq \varepsilon/3$. There is a β_s-neighbourhood V of T in B_{M^s} so that if $S \in V$ then $\tilde{h}(T) - \tilde{h}(S) \in U/3$. Then if $S \in V$,

$$h(S) - h(T) = (h(S) - \tilde{h}(S)) + (\tilde{h}(S) - \tilde{h}(T)) + (\tilde{h}(T) - h(T)) \in U.$$

2.5. <u>Corollary (KAPLANSKY's density theorem)</u>: Let A be a β_{s^*}-dense $*$-subalgebra of the von Neumann algebra M. Then B_A, the unit ball of A, is β_{s^*}-dense in B_M.

<u>Proof</u>: We first show that B_{A^s} is β_s-dense in B_{M^s}. We can assume that A is norm-closed. If $T \in B_{M^s}$ there is a net $(T_\alpha)_{\alpha \in A}$ in A with $T_\alpha \longrightarrow T$ in β_s. We can assume, by I.1.14, that $\{T_\alpha\}$ is bounded. Let h be a continuous function on $\mathbb{R}$ so that $\|h\|_\infty = 1$ and $h = Id_{\mathbb{R}}$ on $[-1,1]$. Then. by 2.4, $h(T_\alpha) \longrightarrow h(T)$ in B_{M^s} for β_s. Now $h(T_\alpha)$ is the uniform limit of polynomials in T_α and so is in A.

To deal with non-self-adjoint operators, we use the following trick: let $\tilde{H}$ be a Hilbert space direct sum $H \oplus H$ and identify operators $\tilde{T}$ on $\tilde{H}$ with 2×2 matrices (T_{ij}) where each $T_{ij} \in L(H)$. Let $\tilde{A}$ (resp. $\tilde{M}$) be the set of operators (T_{ij}) whose components are in A (resp. in M). Then $\tilde{A}$ and $\tilde{M}$ satisfy the conditions of the Corollary. Now if $T \in M$ with $\|T\| \leq 1$, then

$$\tilde{T} := \begin{bmatrix} 0 & T^* \\ T & 0 \end{bmatrix}$$

is in $\tilde{M}^s$ and $\|\tilde{T}\| \leq 1$. Hence there is a net $(T^\alpha) = (T^\alpha_{ij})$ in $B_{\tilde{A}^s}$ so that $T^\alpha \longrightarrow \tilde{T}$ in β_s. Then $T^\alpha_{21} \in B_A$ and $T^\alpha_{21} \longrightarrow T$ in β_{s^*}.

We now present one of the most important results of this Chapter: the result that (M, β_{s^*}) is a Mackey space.

We require the following characterisation of normal forms on a von Neumann algebra. A fairly accessible proof can be found in RINGROSE [13].

2.6. Proposition: Let M be a von Neumann algebra, f a form in M^*. Then f is β_σ-continuous (i.e. $f \in M_*$) if and only if for each orthogonal family $\{E_\alpha\}$ of projections in M
$f(\sum_\alpha E_\alpha) = \sum_\alpha f(E_\alpha)$.

2.7. Corollary: $f \in M^*$ is β_σ-continuous if and only if $f|_{M_1}$ is β_σ-continuous for each commutative von Neumann subalgebra M_1 of M.

2.8. Proposition: Let M be a von Neumann algebra, K a subset of M_*. Then the following are equivalent:

1) K is relatively $\sigma(M_*,M)$-compact;

2) for each commutative von Neumann subalgebra M_1 of M, $K|_{M_1}$ is $\sigma(M_{1*},M_1)$-compact.

Proof: 1) $\Longrightarrow$ 2) is trivial.

2) $\Longrightarrow$ 1): Firstly, if K satisfies 2) then K is pointwise bounded on M^s and so pointwise bounded on M. Hence, by the principle of uniform boundedness, K is norm-bounded and so relatively weakly compact in M^*. Thus it will suffice to show that the $\sigma(M^*,M)$-closure of K lies in M_*. Suppose that f lies in this closure. Then $f|_{M_1}$ lies in the weak closure of $K|_{M_1}$

for each commutative von Neumann subalgebra M_1 of M. Thus f is normal (2.7).

2.9. Corollary: $\tau(M,M_*)$ is the finest locally convex topology on M which is coarser than $\tau(M_1,M_{1*})$ on each commutative von Neumann subalgebra M_1 of M.

2.10. Lemma: Let (T_n) be a sequence in B_{M^s} with $T_n \longrightarrow 0$ in β_s. Then for each $\delta > 0$ there is a sequence (E_n) of projections in M so that $E_n \longrightarrow 1$ in β_s and $\|T_nE_n\| \leq \delta$ for each n.

Proof: Let $E_n := \chi_{[-\delta,\delta]}(T_n)$. Then since

$$T_n^2 \geq \delta^2(I - E_n) \geq 0$$

$I - E_n \longrightarrow 0$ in β_s. Also $\|T_nE_n\| \leq \delta$

2.11. Proposition: Let M be a von Neumann algebra of countable type. Then $\beta_{s^*} = \tau(M,M_*)$.

Proof: We shall work in the real vector space M^s and show that $\beta_s = \tau(M^s,M_*^s)$ there. Since M is of countable type, B_{M_s} is β_s-metrisable and so it will suffice to show that if (T_n) is a sequence in B_{M^s} so that $T_n \longrightarrow 0$ in β_s then $T_n \longrightarrow 0$ uniformly on each weakly compact subset K of M_*^s. If this were not the case, there would be a relatively compact sequence (f_n), $\varepsilon > 0$ and a subsequence of (T_n) (which we assume, for simplicity of notation, to be (T_n) itself) so that $|f_n(T_n)| > \varepsilon$ for each n. Now by EBERLEIN's theorem, (f_n) has a weakly convergent subsequence and so, by passing to a subsequence and by translating, we can assume that $f_n \longrightarrow 0$ weakly. Choose projections $\{E_n\}$ so that $E_n \longrightarrow 1$ in β_s and $\|E_n T_n\| \leq \delta$ for some $\delta > 0$ to be chosen later (2.10). Then

$$|f_j(T_i)| \leq |f_j(E_i T_i)| + |f_j((I - E_i)T_i)|$$
$$\leq 2M\delta + |f_j((I - E_i)T_i)|$$

for each $i,j \in \mathbb{N}$ where $M := \sup\{\|f\| : f \in K\}$.

Now $f_j \longrightarrow 0$ pointwise on the $\sigma(\mathcal{M}^s, \mathcal{M}_*^s)$-compact set $B_{\mathcal{M}^s}$ and so, by Baire's theorem, there is a point $T_o \in B_{\mathcal{M}^s}$ so that (f_j) is equicontinuous at T_o i.e. there is a $\sigma(\mathcal{M}^s, \mathcal{M}_*^s)$-neighbourhood U of zero in $\mathcal{M}^s$ so that

$$|f_j(T) - f_j(T_o)| < \delta$$

for each $T \in V := (T_o + U) \cap B_{\mathcal{M}^s}$. Choose $J \in \mathbb{N}$ so that $|f_j(T_o)| < \delta$ if $j \geq J$. Then

$$|f_j(T)| < 3\delta$$

for $j \geq J$, $T \in V$. Now if

$$S_i := E_i T_o E_i + (I - E_i)T_i$$

then $S_i \in B_{\mathcal{M}^s}$ and $S_i \longrightarrow T_o$. Hence there is an $I \in \mathbb{N}$ so that S_i and $E_i T_o E_i$ are in V for $i \geq I$. Then

$$|f_j((I - E_i)T_i| \leq |f_j(S_i)| + |f_j(E_i T_o E_i)| \leq 6\delta$$

for $i \geq I$, $j \geq J$. Then

$$|f_j(T_i)| \leq (2M + 6)\delta$$

for $i \geq I$, $j \geq J$ and for small δ this contradicts the inequality $|f_n(T_n)| > \varepsilon$.

2.12. Remark: It follows from a characterisation of relatively weakly compact subsets of M_* due to AKEMANN (see [3] or AARNES [2]) that 2.11 holds without the assumption that M be of countable type. However, the extension to general von Neumann algebras requires some more technical results. Nevertheless, in view of this fact, we shall drop the assumption of countable type in the formulation of the following consequences of 2.11.

2.13. Proposition: Let M be a von Neumann algebra, F a locally convex space. Then a linear mapping Φ from M into F is β_{s^*}-continuous if and only if its restriction to each commutative von Neumann subalgebra of M is β_s-continuous.

Proof: Since (M, β_{s^*}) is a Mackey space, we can reduce to the case when $F = \mathbb{C}$ and this is precisely 2.7.

2.13 is equivalent to the fact that β_{s^*} is the finest locally convex topology on M which is coarser than β_s on each commutativ von Neumann subalgebra M_1 of M (cf. 2.9).

2.14. Proposition: If M is a von Neumann algebra, then (M, β_{s^*}) has the Banach-Steinhaus property.

Proof: By 2.13 we can reduce this to the case where M is commutative and this is III.1.12.

2.15. Corollary: If M is a von Neumann algebra then M_* is $\sigma(M_*,M)$-sequentially complete.

2.16. Proposition: Let M and M_1 be von Neumann algebras, Φ a linear mapping from M into M_1. Then Φ is β_σ-continuous if and only if it is β_{s^*}-continuous. If, in addition, Φ is a $*$-morphism, then these conditions are equivalent to its continuity for β_s.

Proof: If Φ is β_σ-continuous, then it is continuous for the Mackey topologies $\tau(M,M_*)$ and $\tau(M_1,M_{1*})$ i.e. for the β_{s^*}-topologies.
On the other hand, if it is Mackey continuous, then its adjoint (which maps M_{1*} into M_*) is weakly continuous and so norm continuous. Hence it maps norm compact sets into norm compact sets and so the result follows from 2.3.5).

If Φ is β_s-continuous, then it is β_{s^*}-continuous (proof as for β_σ-continuity). On the other hand, if Φ is star-preserving, then $\Phi(M^s) \subseteq M_1^s$ and $\beta_s = \beta_{s^*}$ on M^s and M_1^s. Hence if Φ is β_{s^*}-continuous, then it is β_s-continuous on M^s and the result follows by considering real and imaginary parts.

2.17. Proposition: If M is a von Neumann algebra, then a linear mapping Φ from (M,β_{s^*}) into a separable Fréchet space is continuous if and only if it has a closed graph.

Proof: By 2.13, it is sufficient to show that the restriction of Φ to each commutative von Neumann subalgebra M_1 of M is β_s-continuous. But $\Phi|_{M_1}$ also has a closed graph and so is continuous by III.1.13.

As mentioned in the introduction, von Neumann defined two locally topologies, the ultraweak and ultrastrong topologies, on L(H) (and so on any von Neumann algebra) as follows: if $X := (x_1,x_2,...)$ and $Y := (y_1,y_2,...)$ are sequences of elements in H so that $\Sigma \|x_n\|^2 < \infty$ and $\Sigma \|y_n\|^2 < \infty$ then

$$p_X : T \longmapsto \Sigma \|Tx_n\|^2$$

$$p_{X,Y} : T \longmapsto \Sigma |(Tx_n|y_n)|$$

are seminorms on L(H) and the family of all such seminorms $\{p_X\}$ (resp. $\{p_{X,Y}\}$) defines a locally convex topology on L(H) - the ultrastrong (resp. ultraweak) topology. We denote these topologies by us and $u\sigma$. Then $\tau_\sigma \subseteq u\sigma$ and $u\sigma = \tau_\sigma$ on $B_{L(H)}$. Similary, $\tau_s \subseteq$ us and us $= \tau_s$ on $B_{L(H)}$ (see DIXMIER [8], p. 34).

One can also define a symmetric form us* of us (by adding the seminorms $T \longmapsto p_X(T^*)$) and us* $\supseteq \tau_{s^*}$ with equality on the unit ball. Thus β_σ, β_s and β_{s^*} are stronger than $u\sigma$, us and us* respectively. In general, they are strictly stronger as the following Proposition shows:

2.18. Proposition: Let M be an infinite dimensional von Neumann algebra. Then $u\sigma$, us and us^* are strictly weaker than β_σ, β_s and β_{s^*} respectively.

Proof: The topology $u\sigma$ on M is exactly $\sigma(M,M_*)$ and β_σ is the topology of uniform convergence on the norm compact subsets of M_*. Hence $u\sigma = \beta_\sigma$ if and only if the norm compact subsets of M_* are finite dimensional and this is the case precisely when M_* (and hence also M) is finite dimensional.
For the second part, it suffices to show that β_s is strictly finer than us on M^s. Since M is infinite-dimensional, there is a mutually orthogonal sequence (E_n) of non-zero projections in M^s. Let $S := \{n^{1/2}E_n\}$. Then we can claim that 0 lies in the us-closure of S since for each X, $p_X(n^{1/2}E_n) \le 1$ for some n. For if this were not the case, there is an $X = (x_i)$ so that $p_X(n^{1/2}E_n) > 1$ i.e. $p_X(E_n) > n^{-1/2}$. But this contradicts the additivity of the form f_X in M_* defined by X (i.e. $\Sigma\,\omega_{x_i,y_i}$ in the notation of DIXMIER [8]) since then $\Sigma\, f_X(E_n) \ge \Sigma\, 1/n$.
We shall show that 0 does not lie in the β_s-closure.
For each n let x_n be a unit vector in H so that $E_n x_n = x_n$ and define f_n to be the form

$$T \longmapsto n^{-1/2}(Tx_n|x_n).$$

Then $f_n \in M_*$ and $K := \{f_n\}$ is relatively compact in $(M_*^s, \|\ \|)$.

Hence

$$U := \{T \in M^s : |f_n(T)| < 1 \text{ for each } n\}$$

is a β_s-neighbourhood of zero in M^s. But $S \cap U = \emptyset$ since $f_n(n^{1/2}E_n) = 1$ for each n.

2.19. W*-algebras: A W*-algebra is a C*-algebra $(A, \| \|)$ which is, as a Banach space, the dual of a Banach space A_*.
Two W*-algebras A and B are isomorphic if there is a C*-algebra isomorphism between them which is bicontinuous for the corresponding weak topologies $\sigma(A,A_*)$ and $\sigma(B,B_*)$. We denote by T the set of positive elements of A_*. Then, on A, one can define the following locally convex topologies:

$\sigma(A,A_*)$ - the weak topology;

$s(A,A_*)$ - the topology defined by the seminorms

$$x \longmapsto \{\phi(x^*x)\}^{1/2} \qquad (\phi \in T);$$

$s^*(A,A_*)$ - the topology defined by the seminorms

$$x \longmapsto \{\phi(x^*x)\}^{1/2}$$
$$x \longmapsto \{\phi(xx^*)\}^{1/2} \qquad (\phi \in T).$$

(see SAKAI [14], §§ 1.7 and 1.8).

Now it is clear that every von Neumann algebra M is a W*-algebra (since it is the dual of M_*). On the other hand, one has the following result of SAKAI ([14], Theorem 1.16.7):

every W*-algebra is isomorphic to a von Neumann algebra.
One then has the following relationship between the topologies

of a W^*-algebra A and the topologies that it possesses by virtue of being a von Neumann algebra:

$$u\sigma = \sigma(A,A_*); \quad us = s(A,A_*); \quad us^* = s^*(A,A_*).$$

Thus all our result can be carried over to the context of W^*-algebras.

2.20. Remark: There are two alternative methods of defining mixed topologies on W^*-algebras which we now describe briefly. If M is a von Neumann algebra, a (two-sided, norm-closed) ideal $I \subseteq M$ is essential if I^0, the annihilator

$$\{x \in M : xI = \{0\}\}$$

of I, is $\{0\}$. The family of all essential ideals is denoted by E. E is closed under the operations of sum and intersection i.e. $I,J \in E \Longrightarrow I + J$ and $I \cap J$ are in E. If I is an essential ideal in M, we define the strict topology β_I on M to be that locally convex topology generated by the seminorms:

$$x \longmapsto \|ax\| \quad (a \in I), \quad x \longmapsto \|xb\| \quad (b \in I).$$

Then we define $\bar{\beta}$ to be the inductive locally convex topology generated by $\{\beta_I : I \in E\}$ i.e. it is the finest locally convex topology on M, which is coarser than each β_I, $I \in E$. The topology $\bar{\beta}$ was introduced by HENRY and TAYLOR in [11] and their main result can be summarised as follows:

$\bar{\beta} = \beta_{s^*}$ if and only if $\bar{\beta}$ is a topology of the duality (M,M_*) and this is the case if M is a countable decomposable type I algebra.

Hence, in this case, $\bar{\beta}$ coincides with β_{s^*}.

The second approach uses the theory of non-commutative integration which has been developed by SEGAL. With this notion, one can define a mixed topology on a W*-algebra using methods which are exact analogues of those used in Chapter III.
We recall briefly this theory and refer the reader to NELSON [12] for details. Let M be a W*-algebra. A normal faithful semi-finite trace on M is a mapping $\tau : M^+ \longmapsto [0,\infty]$ (M^+ denotes the set of positive elements of M^s) which satisfies the condition:

1) $\tau(x+y) = \tau(x) + \tau(y) \qquad (x,y \in M^+)$;

2) $\tau(\lambda x) = \lambda\tau(x) \qquad (x \in M^+, \lambda \in [0,\infty[)$;

3) $\tau(x^*x) = \tau(xx^*) \qquad (x \in M^+)$;

4) if (x_α) is an increasing net in M^+ and $x = \sup\{x_\alpha\}$ (we write $x_\alpha \uparrow x$ to describe this situation briefly) then $\tau(x_\alpha) \longrightarrow \tau(x)$;

5) if $\tau(x) = 0$ then $x = 0$;

6) if $x \in M^+$ there is a net (x_α) in M_0^+ (the set of those y in M^+ with $\tau(y) < \infty$) so that $x_\alpha \uparrow x$.

Note that the existence of such a trace is equivalent to the assumption that M be semi-finite (see SAKAI [14], 2.5.4 and 2.5.7). If M is finite in the sense that τ takes finite values on M^+ then we can define analogues of the L^p-spaces as follows:

$$||x||_p := (\tau(|x|^p))^{1/p}$$

(where $|x|$ is the absolute value of $x \in M$) is a norm on M and we denote by M_p its completion. Then $(M, \| \|, \| \|_p)$ is a Saks space. We denote by β_1 the mixed topology $\gamma[\| \|, \| \|_p]$.

In the general case (i.e. where τ is not necessarily finite valued) we can regard M as the Banach space projective limit of the ideals $\{eMe\}$ as e ranges over the projections in M_o^+. We can then provide M with a Saks space structure as the Saks space projective limit of $\{(eMe, \| \|, \| \|_p)\}$ and we denote the corresponding mixed topology by β_1. We conjecture that, in this case, β_1 and β_{s^*} coincide once again.

We conclude this chapter with some remarks on von Neumann algebras.

2.21. Remarks: I. AARNES [2] has given an interesting order-theoretical characterisation of the topology β_{s^*}. Recall that M^s has a natural ordering and so we can define the notion of order convergence for a net in M^s and so in M (by considering the real and imaginary parts). Then β_{s^*} is the finest locally convex topology τ on M for which order convergence is stronger than τ-convergence.

II. We remark that the finiteness classification of von Neumann algebras can be expressed elegantly in terms of the mixed topologies as follows:

1) M is finite if and only if $\beta_s = \beta_{s^*}$;

2) M is semi-finite if and only if it is the inductive limit (in the sense of Banach spaces) of a directed set of weakly closed ideals for which $\beta_s = \beta_{s^*}$;

3) M is purely infinite if and only if $\beta_s \neq \beta_{s^*}$ on each weakly closed (non-zero) ideal of M.
(cf. SAKAI [14], Th. 2.5.6).

III. 2.14 can be strengthened to give the following result of VITALI-HAHN-SAKS type. Let (T_n) be a sequence of β_{s^*}-continuous linear operators from M into a complete locally convex space F so that for each projection P in M, $\lim T_n(P)$ exists. Then $\{T_n\}$ is β_{s^*}-equicontinuous and there is a β_{s^*}-continuous linear operator T from M into F so that $T_n(P) \longrightarrow T(P)$ for each projection. For, using 2.11 and 2.7, we can reduce successively to the case where $F = \mathbb{C}$ and M is commutative i.e. to the classical Vitali-Hahn-Saks theorem (cf. AARNES [1]).

IV. It follows easily from the results of Chapter I that if M is a von Neumann algebra, then M is barrelled, bornological or nuclear for any of the β-topologies if and only if it is finite dimensional. Also if $(M,\| \|)$ is separable, then, by 2.17, $\beta_{s^*} = \| \|$ and this also implies that M is finite dimensional. Also, if M is semi-Montel under β_s or β_{s^*} then these topologies coincide with β_σ and this implies, once again, that M is finite dimensional.

V. If $(A,\| \|)$ is a C*-algebra, then $(A,\| \|,\sigma(A,A'))$ is a Saks space. We denote its completion by $(\hat{A},\| \|,\hat{\sigma})$ (i.e. $\hat{A}$ is the bidual of A). Then involution is a Saks space morphism and so extends to an involution on $\hat{A}$. Similarly, multiplication can be extended (in two steps) to $\hat{A}$. Then $(\hat{A}, \| \ \|)$ is a C^*-algebra and, since it is a dual space, even a W^*-algebra. $\hat{A}$ is the <u>enveloping W^*-algebra</u> of A. It has the universal property that every C^*-morphism from A into a W^*-algebra can be extended to a W^*-algebra morphism on $\hat{A}$.

II.3. NOTES

A readable introduction to von Neumann algebras has been given by RINGROSE in [13].

Some results from this chapter were announced in COOPER [5]. The representation of the predual of L(H) as N(H) is due to DIXMIER (who, in fact, used the ultraweak topology). The representation theorem for commutative algebras can be found in [8], § I.7 or in [14], § 1.18. Full details on spectral theory can be found in DUNFORD and SCHWARTZ [9]. The claim that the functional calculus defines a β_s-continuous $*$-algebra homomorphism follows from formula X.2.8(ii) there. The idea of the proof of 2.5 (using 2.4) is sketched by SHIELDS in [15] (using a different topology). Proposition 2.6 is due to DIXMIER (for positive forms) and SAKAI. The result 2.8 - 2.11 are based on papers [3] and [2] of AKEMANN and AARNES. In particular, the proof of 2.11 is taken from [3]. 2.15 is a result of SAKAI.

For a full discussion of the topologies discussed in 2.18 and 2.19 see [8], § I.3 and [14], §§ 1.7, 1.8. The difficult part of 2.18 is due to YEADON [16].

We mention the following related articles:

DAUNS [7] examines the problem of tensor problems for W*-algebras;

GUICHARDET [10] studies W*-algebras from a category theoretical point of view;

DARST [6] considers a theorem of Vitali-Hahn-Saks type for the norm dual of M.

REFERENCES FOR CHAPTER IV.

[1] J.F. AARNES The Vitali-Hahn-Saks theorem for von Neumann algebras, Math. Scand. 18 (1966) 87-92.

[2] On the Mackey topology for a von Neumann algebra, Math. Scand. 22 (1968) 87-102.

[3] C.A. AKEMANN The dual space of an operator algebra, Trans. Amer. Math. Soc. 126 (1967) 286-302.

[4] Sequential convergence in the dual of a W*-algebra, Comm. Math. Physic 7 (1968) 222-224.

[5] J.B. COOPER Topologies sur l'espace des opérateurs dans l'espace hilbertien, C.R. Acad. Sci. Paris A 276 (1973) 1509-1511.

[6] R.B. DARST On a theorem of Nykodym with applications to weak convergence in von Neumann algebras, Pac. J. Math. 23 (1967) 473-477.

[7] J. DAUNS Categorical W*-tensor products, Trans. Amer. Math. Soc. 166 (1972) 439-456.

[8] J. DIXMIER Les algèbres d'opérateurs dans l'espace hilbertien (Paris, 1969).

[9] N. DUNFORD, J. SCHWARTZ Linear operators. Part II (New York, 1963).

[10] A. GUICHARDET Sur la catégorie des algèbres de von Neumann, Bull. Sci. Math. (2) 90 (1966) 41-46.

[11] J. HENRY, D.C. TAYLOR The $\bar{\beta}$-topology for W*-algebras, Pac. Jour. Math. 60 (1975) 123-139.

[12] E. NELSON Notes on non-commutative integration, Jour. Func. Anal. 15 (1974) 103-116.

[13] J.R. RINGROSE Lectures on the trace in a finite von Neumann algebra (in "Lectures on Operator Algebras", pp. 313-354, Springer Lecture Notes 247, 1972).

[14] S. SAKAI C^*-algebras and W^*-algebras (Berlin, 1971)

[15] P.C. SHIELDS A new topology for von Neumann algebras, Bull. Amer. Math. Soc. 65 (1959) 267-269.

[16] F.J. YEADON A note on the Mackey topology of a von Neumann algebra, J. Math. Anal. Appl. 45 (1974) 721-722.

CHAPTER V - SPACES OF BOUNDED HOLOMORPHIC FUNCTIONS

Introduction: In this chapter we consider the algebra $H^\infty(G)$ of bounded, holomorphic functions on an open domain G in $\mathbb{C}$. This algebra has a natural Banach algebra structure (with the supremum norm). Following BUCK [10] we list some unpleasant features of this Banach algebra (for the special case where G is the open unit disc U):

1) The polynomials are not dense in $H^\infty(U)$ and, in fact, $H^\infty(U)$ is not separable;

2) principal ideals in $H^\infty(U)$ are not necessarily closed and closed ideals are not necessarily principal;

3) there are maximal ideals of $H^\infty(U)$ which are not determined by points of U.

In 1957 BUCK introduced the strict topology β on H^∞ and showed that it possessed several attractive properties. It was then studied in more detail by RUBEL, RYFF and SHIELDS. In this chapter we show that β is an example of a mixed topology. In section 1, we deduce the basic properties of β using the theory of Chapter I. In V.2 we specialise to the case where G is the open unit disc. We introduce a new, finer mixed topology β_1 which seems to be the appropriate topology for certain applications and study tensor products and vector-valued functions. In V.3 we consider $(H^\infty(U),\beta)$ as a Saks algebra and obtain a characterisation of its closed ideals. We conclude with an application of the theory developed to operators in Hilbert space.

Most of the results in this Chapter are known. However, the methods are new.

V.1. MIXED TOPOLOGIES ON H^∞

In this section we denote by G an open domain in $\mathbb{C}$ which supports a non-constant bounded holomorphis function. Then the spece $H^\infty(G)$ of bounded holomorphic functions on G separates G. On $H^\infty(G)$ we consider the following structures:

$\| \ \|$ - the supremum norm;

τ_p - the topology of pointwise convergence on G;

τ_K - the topology of uniform convergence on the compact subsets of G.

Then $(H^\infty, \| \ \|, \tau_p)$ and $(H^\infty, \| \ \|, \tau_K)$ are Saks spaces and, since the unit ball of H^∞ is a normal family of functions, τ_p and τ_K coincide there. Hence

$$\gamma[\| \ \|, \tau_p] = \gamma[\| \ \|, \tau_K]$$

and we denote this topology by β. We list now some simple properties of β which follow immediately from the theory of Ch. I.

1.1. <u>Proposition</u>: 1) (H^∞, β) is a complete locally convex space;
2) the β-bounded sets of H^∞ are precisely the norm-bounded sets;
3) a sequence (x_n) in H^∞ converges to zero if and only if it is norm bounded and converges pointwise to zero;

4) (H^∞,β) is semi-Montel and so is the dual of the Banach space $(H^\infty,\beta)'$, with the topology of compact convergence;

5) a linear mapping from H^∞ into a locally convex space is β-continuous if and only if its restriction to B_{H^∞}, the unit ball of H^∞, is τ_p-continuous (and hence if and only if it is β-sequentially continuous).

Now (H^∞,β) is the dual of the Banach space $F := (H^\infty,\beta)'$ and F is separable (since it is a subspace of the separable space $C(B_{B^\infty})$ of continuous functions on the compact metrisable space B_{H^∞}). Hence the following properties follow easily from I.4.1-3:

a subset A of H^∞ is β-closed if and only if it is sequentially closed (for A is closed if $A \cap nB_{H^\infty}$ is closed for each n and the latter space is metrisable);

β is the finest topology on H^∞ with convergent sequences those described in 1.1.3) (follows immediately from I.4.2(c) and the fact that each nB_{H^∞} is metrisable).

We can regard $(H^\infty(G),\|\ \|,\tau_K)$ as a Saks subspace of $(C^\infty(G),\|\ \|.\tau_K)$. By I.4.4, (H^∞,β) is a locally convex subspace of $(C^\infty(G),\beta)$ and so β is defined by the seminorms

$$p_\phi : x \longmapsto \|x\phi\|$$

as ϕ runs through the space $C_o(G)$. Hence (H^∞,β) coincides with the space $\beta(G)$ studied by RUBEL, RYFF and SHIELDS ([30],[31]). In particular, (H^∞,β) is a topological algebra. Note, however, that inversion is not β-continuous.

We can also regard $H^\infty(G)$ as a Saks subspace of $(L^\infty(G), \|\ \|, \sigma)$ (G is provided with planar Lebesgue measure) and once again (H^∞, β) is a locally convex subspace of $(L^\infty(G), \beta_\sigma)$. Hence by III.1.1 we have

1.2. Proposition: $(H^\infty(G), \beta)'$ is naturally isometric to the quotient spaces $L^1(G)/N_1$ and $M(G)/N_2$ where N_1 and N_2 are the polars of H^∞ in the appropriate dual spaces.

We note that the above results show that if $f \in (H^\infty, \beta)'$ then

$$\|f\| = \inf\{\|\mu\| : \mu \in M(G),\ \pi(\mu) = f\}$$

where π is the natural projection from M(G) onto the dual of H^∞. This infimum is attained if and only if f is equivalent to a measure of the form $a\nu$ where $a \in \mathbb{C}$ and ν is a positive measure (see RUBEL and SHIELDS [31], Prop. 2.10).

Note that by identifying $L^1(G)$ with a subspace of M(G) in the usual way, we can regard this result as one on "balayage":

if $\mu \in M(G)$, $\varepsilon > 0$ then there is an absolutely continuous measure (i.e. absolutely continuous with respect to Lebesgue measure) ν in M(G) so that

$$\mu \sim \nu \qquad \text{and} \qquad \|\nu\| \leq \|\mu\| + \varepsilon$$

(where $\mu \sim \nu$ means that $\mu - \nu$ vanishes on H^∞).
In fact, the result is true with the ε-term omitted (cf. RUBEL and SHIELDS [31], Prop. 4.1).

1.3. <u>Definition</u>: Let S be a subset of G. S is <u>dominating</u> if, for each $x \in H^{\infty}(G)$,

$$\sup\{|x(\lambda)| : \lambda \in S\} = \sup\{|x(\lambda)| : \lambda \in G\}$$

i.e. if the restriction mapping $\rho_S : x \longmapsto x|_S$ is an isometry from $H^{\infty}(G)$ into $\ell^{\infty}(S)$.

1.4. <u>Lemma</u>: If S is a countable, dominating set then $\rho_S(H^{\infty})$ is β-closed in $\ell^{\infty}(S)$.

<u>Proof</u>: By I.4.3 and the fact that the unit ball of $\ell^{\infty}(S)$ is β-metrisable, it is sufficient to show that $\rho_S(H^{\infty})$ is sequentially closed. Let (x_n) be a sequence in H^{∞} so that $\rho_S(x_n)$ is β-convergent in $\ell^{\infty}(S)$ to y. Then (x_n) contains a β-convergent subsequence (x_{n_k}) and it is clear that $y = \rho_S(\lim x_{n_k})$.

It follows from this Lemma that ρ_S is an isomorphism from $(H^{\infty}(G),\beta)$ onto a closed subspace of $(\ell^{\infty}(S),\beta)$ (cf. I.4.32).

1.5. <u>Proposition</u>: Let S be a subset of G. Then the following are equivalent:

1) S is dominating;

2) for every $\mu \in M(G)$ there is a ν in $M(S)$ so that $\mu \sim \nu$ (i.e. $\mu - \nu$ vanishes on H^{∞}).

<u>Proof</u>: 2) $\Longrightarrow$ 1): if S were not dominating, we could find a $\lambda_o \in G \setminus S$ and $x \in H^{\infty}(G)$ so that

$$x(\lambda_o) = 1, \quad |x(\lambda)| \le k \quad \text{for } \lambda \in S \ (k < 1).$$

If S satisfies condition 2) there is a $\nu \in M(S)$ so that $\nu \sim \delta_{\lambda_o}$ (Dirac measure at λ_o). Now for each $n \in \mathbb{N}$,

$$1 = \int x^n \, \delta_{\lambda_o} = \int x^n d\nu$$

and so $1 \le k^n ||\nu||$ which is impossible.

$1 \Longrightarrow 2)$: by going over to a countable dense subset, we can suppose that S is countable. The result then follows immediately from 1.4 and duality.

V.2. THE MIXED TOPOLOGIES ON $H^\infty(U)$

In this section, we specialise to function spaces on U, the open unit disc $\{\lambda : |\lambda| < 1\}$. We denote by $\overline{U}$ the closed unit disc and by ∂U the boundary $\overline{U} \setminus U$ of U. We recall some notation and results on Hardy spaces -see HOFFMAN [21], Ch. 3,4 for a detailed discussion.

We denote by $H^p(U)$ (or simply by H^p) $(1 \le p < \infty)$ the space of functions x which are analytic on U and are such that the norm

$$||x||_p := \sup_{0<r<1} \left(\int_0^{2\pi} |x(re^{i\theta})|^p d\theta \right)^{1/p}$$

is finite. $(H^p, || \ ||_p)$ is a Banach space.
Then we have $H^\infty \subseteq H^p \subseteq H^1$ $(1 < p < \infty)$ and the natural inclusion mappings are continuous (when H^∞ has the supremum norm).

Each function x in H^1 has a non-tangential boundary function, which is in $L^1(\partial U)$. More precisely, there exists a function $\tilde{x} \varepsilon L^1(\partial U)$ so that, for almost every $e^{i\theta}$ in ∂U,

$x(\lambda) \longrightarrow \tilde{x}(e^{i\theta})$ as $\lambda \longrightarrow e^{i\theta}$ non-tangentially

(in particular, radially). The mapping $x \longmapsto \tilde{x}$ is an isometry from $H^p(U)$ onto a closed subspace of $L^p(\partial U)$. The range of this mapping is the space of those $y \varepsilon L^p(\partial U)$ such that

$$\int_0^{2\pi} y(e^{i\theta})e^{in\theta}d\theta = 0 \quad (n = 1,2,\ldots)$$

i.e. the non-negative Fourier coefficient of y vanish. If $y \varepsilon L^p(\partial U)$ satisfies these conditions, then $y = \tilde{x}$ where $x \varepsilon H^p(U)$ can be recovered from y via the Poisson integral

$$x(re^{it}) = \frac{1}{2\pi}\int_0^{2\pi} \frac{y(e^{i\theta})(1-r^2)}{1 - 2r\cos(t-\theta) + r^2}\, d\theta$$

2.1. <u>Proposition</u>: 1) $(H^\infty(U),\beta)$ is a topological subspace of $(L^\infty(\partial U),\beta_\sigma)$;

2) $(H^\infty(U),\beta)'$ is naturally isomorphic to the quotient space $L^1(\partial U)/H^1_o$ where H^1_o denotes the space of functions in $H^1(U)$ which vanish at the origin.

<u>Proof</u>: 1) follows from I.4.4 and the fact that on the unit ball of H^∞, the topology τ_p agrees with the weak topology $\sigma(L^\infty(\partial U),L^1(\partial U))$.

2) Since the dual of $(L^\infty(\partial U),\beta_\sigma)$ is $L^1(\partial U)$, it suffices to identify the polar of H^∞ in $L^1(\partial U)$ with H^1_o and this is clear.

For the next Proposition, we recall the following definition: a sequence $(x_n)_0^\infty$ in a topological vector space E is a <u>Cesaro basis</u> for E if for each $x \in E$ there is a unique sequence (ξ_n) of scalars so that the partial sums

$$s_n := \sum_{k=0}^{n} \xi_k x_k$$

converge to x in Cesaro mean (i.e. $\lim \frac{1}{n+1}(s_0 + \ldots + s_n) = x$).

2.2. <u>Proposition</u>: 1) The sequence $(z_n)_0^\infty$ is a Cesaro basis for (H^∞, β) where z_n is the function

$$\lambda \longmapsto \lambda^n$$

2) (H^∞, β) is separable;

<u>Proof</u>: 1) If $x \in H$, let $\sum_{0}^{\infty} \xi_n \lambda^n$ be its Taylor expansion. Then if

$$s_n := \sum_{k=0}^{n} \xi_k z_k, \qquad \sigma_n := \frac{1}{n+1}(s_0 + \ldots + s_n)$$

we have $s_n \longrightarrow x$ in τ_K and so $\sigma_n \longrightarrow x$. On the other hand, we have the following kernel representation of σ_n:

$$\sigma_n(e^{it}) = \frac{1}{2\pi} \int_{-\pi}^{\pi} x(e^{i\theta}) K_n(t-\theta) d\theta$$

where K_n is the Fejer kernel

$$t \longmapsto \frac{1}{n} \left[\frac{1 - \cos nt}{1 - \cos t}\right]$$

(cf. HOFFMAN [21], pp.16,17 and Ch. 3).

Then $\|\sigma_n\| \leq \|x\|$ (since K_n is positive and $\frac{1}{2\pi} \int_{-\pi}^{\pi} K_n = 1$).

Hence $\sigma_n \longrightarrow x$ in β by I.1.10.

2) follows immediately from 1).

It is well known that if E is a locally convex space with a basis (x_n) such that the corresponding projection operators

$$p_n : \sum_{k=0}^{\infty} \xi_k x_k \longmapsto \sum_{k=0}^{n} \xi_k x_k$$

are equicontinuous, then E has the approximation property.

The following Lemma is proved similarly.

2.3. <u>Lemma</u>: Let E be a locally convex space with a Cesaro basis (x_n) so that the projection mappings (P_n) are equicontinuous, where P_n maps $\Sigma\, \xi_k x_k$ into the Cesaro mean σ_n. Then E has the approximation property.

2.4. <u>Proposition</u>: (H^∞,β) has the approximation property.

<u>Proof</u>: By 2.3 and I.1.7 it is sufficient to show that the projections (P_n) are equicontinuous for the topology defined by the seminorms $\{p_r : 0 < r < 1\}$ where

$$p_r : x \longmapsto \sup\{|x(re^{i\theta})| : \theta \in [0,2\pi]\}$$

But exactly the same argument as used in the proof of 2.2 shows that $p_r(\sigma_n) \le p_r(x)$ - q.e.d.

Now we introduce a new mixed topology on $H^\infty(U)$. We consider H^∞ as a Saks space with the supremum norm and the auxiliary topology $\tau_{\|\ \|_p}$ induced by the norm $\|\ \|_p$. Note that $\tau_{\|\ \|_p}$ is finer than the topology τ_p of pointwise convergence on U (as one can see, for example, from the Cauchy integral formula) and so the unit ball of H^∞ is $\tau_{\|\ \|_p}$-closed - in fact, $\tau_{\|\ \|_p}$-complete. We define the topology β_1 to be the mixed topology $\gamma[\|\ \|,\tau_{\|\ \|_1}]$.

2.5. <u>Proposition</u>: 1) β_1 is strictly finer than β;

2) $\beta_1 = \gamma[\|\ \|,\tau_{\|\ \|_p}]$ $(1 \leq p < \infty)$;

3) (H^∞,β_1) is a complete locally convex space;

4) $A \subseteq H^\infty$ is β_1-bounded if and only if it is norm-bounded;

5) a sequence (x_n) in H^∞ is β_1-convergent to zero if and only if it is norm-bounded and converges to zero in some H^p-space $(1 \leq p < \infty)$ (or even in measure on ∂U).

<u>Proof</u>: 1) β_1 is finer than β by the above remarks. It is strictly finer since $z_n \longrightarrow 0$ (β) but $z_n \not\longrightarrow 0$ (β_1) ((z_n) the sequence of 2.2.1)).

2.6. <u>Proposition</u>: β and β_1 are topologies of the same dual. Hence (H^∞,β_1) is semi-reflexive.

<u>Proof</u>: By I.1.18(ii) it suffices to show that if $f \in (H^\infty,\beta_1)'$, $\varepsilon > 0$, then there is an $\tilde{f} \in (H^\infty,\beta)'$ so that $\|f - \tilde{f}\| \leq \varepsilon$.

By the same result, we can write $f = f_1 + f_2$ where f_1 is continuous on H^∞, with the norm induced from $L^1(\partial U)$, and $\|f_2\| \leq \varepsilon$. By the Hahn-Banach theorem, f_1 is representable in the form

$$x \longmapsto \int_{\partial U} xy$$

with $y \in L^\infty(\partial U)$ and this form is β-continuous (2.1).

2.7. <u>Corollary</u>: (H^∞, β) is not a Mackey space.

We now consider vector-valued functions. We recall the notion of a holomorphic function with values in a locally convex space. If G is an open subset of $\mathbb{C}$, E a locally convex space, a mapping $x : G \longrightarrow E$ is <u>holomorphic</u> if for each $\lambda_o \in G$ the limit

$$\lim_{\lambda \to \lambda_o} \frac{x(\lambda) - x(\lambda_o)}{\lambda - \lambda_o}$$

exists in E. The following result is basic (cf. GROTHENDIECK [18]).

2.8. <u>Proposition</u>: Let x be a function from G into the complete locally convex space E. Then the following are equivalent:

1) x is holomorphic;

2) for each $f \in E'$, the complex-valued function $f \circ x$ is holomorphic;

3) x is (weakly) continuous and $\int_\Gamma x(\lambda) d\lambda = 0$ for each closed, simple, nullhomotopic, rectifiable curve Γ in G;

4) for each $\lambda_o \in G$ there is a neighbourhood V of λ_o in G

and a closed, absolutely convex bounded subset B of E so that $x(V) \subseteq E_B$, the subspace spanned by B, and x is holomorphic as a mapping from V into the Banach space $(E_B, \|\ \|_B)$.

Now let $(E, \|\ \|, \tau)$ be a complete Saks space, x a norm bounded function from U into E. Then the following are equivalent:

1) x is holomorphic as a function with values in $(E, \|\ \|)$;

2) x is holomorphic as a function with values in (E, γ);

3) x is holomorphic as a function with values in (E, τ);

4) for each $f \in (E, \tau)'$, $f \circ x$ is holomorphic;

5) x is τ-continuous (or even $\sigma(E, E'_\tau)$-continuous and $\int_\Gamma x(\lambda) d\lambda = 0$ for each closed, simple, rectifiable curve Γ in U.

The equivalence of 1) and 2) follows from 2.8.4) since the norm bounded and the γ-bounded subsets coincide. The equivalence of 2) and 3) follows trivially from the fact that γ and τ coincide on the range of x. 3) and 4) are equivalent by 2.8.2) (despite the fact that (E, τ) is not necessarily complete - we can work in the completion).

We denote by $H^\infty(U;E)$ the space of functions which satisfy one (and hence all) of the conditions 1) - 5). On $H^\infty(U;E)$ we consider the structures:

$\|\ \|_E$ - the supremum norm $x \longmapsto \sup\{\|x(\lambda)\| : \lambda \in U\}$

τ_E - the topology of uniform convergence (with respect to τ) on the compact sets of U.

Then $(H^\infty(U;E), \| \|_E, \tau_E)$ is a complete Saks space. We denote by β_E the associated mixed topology $\gamma[\| \|_E, \tau_E]$. Then $(H^\infty(U;E), \beta_E)$ satisfies the appropriate forms of properties 1),2),3) and 5) (without the last paranthesis) of Prop. 1.1.

We shall now give a tensor product representation of $H^\infty(U;E)$ similar to that given in § II.4 for space of vector-valued continuous functions. For this result we require the following fact (see GROTHENDIECK [18]): if E is a complete locally convex space and $x : U \longrightarrow E$ is analytic, then there is a sequence (a_n) in E so that

$$x(\lambda) = \sum_{n=0}^{\infty} a_n \lambda^n \qquad (\lambda \in U).$$

(in fact, a_n is given by the formula $a_n = \frac{1}{n!} x^{(n)}(0)$). The convergence is uniform on the compact subsets of U.

If $x \in H^\infty(U;E)$ we define, just as for the case of scalar-valued functions, the functions (s_n) and (σ_n) by

$$s_n : \lambda \longmapsto \sum_{k=0}^{n} a_k \lambda^k$$

$$\sigma_n := \frac{1}{n+1} (s_0 + \ldots + s_n).$$

2.9. Proposition: If $(E, \| \|, \tau)$ is a complete Saks space and $x \in H^\infty(U;E)$, then $\sigma_n \longrightarrow x$ in β_E.

Proof: We have already seen that $\sigma_n \longrightarrow x$ in τ_E. Hence it suffices to show that $\{\sigma_n\}$ is bounded i.e. that $\bigcup_{n \in \mathbb{N}} \bigcup_{\lambda \in U} \sigma_n(\lambda)$

is bounded in E. By the uniform boundedness theorem, it suffices to show that for each $f \in E'_\gamma$, $\bigcup_{n \in \mathbb{N}} \bigcup_{\lambda \in U} f \circ \sigma_n(\lambda)$ is bounded in $\mathbb{C}$ i.e. we can reduce to the scalar case which we have already treated in 2.2.

2.10. Corollary: If we identify the algebraic tensor product $H^\infty(U) \otimes E$ with a subspace of $H^\infty(U;E)$ under the natural (vector space) isomorphism

$$\Sigma x_k \otimes a_k \longmapsto (\lambda \longmapsto \Sigma x_k(\lambda) a_k)$$

then $H^\infty(U) \otimes E$ is β_E-dense in $H^\infty(U;E)$.

2.11. Proposition: There is a natural (Saks space) isomorphism $H^\infty(U;E) = H^\infty(U) \hat{\hat{\otimes}}_\gamma E$.

Proof: Having identified $H^\infty(U) \otimes E$ with a dense subspace of $H^\infty(U;E)$, we need only show that the Saks space structure induced on $H^\infty(U) \otimes E$ from $H^\infty(U;E)$ coincides with the tensor product structure. This is standard.

We finish this section by quoting two important Propositions. The first states that the dual of $(H^\infty(U), \beta)$ is sequentially complete and is due to MOONEY who thus settled a long-standing conjecture. The second is a characterisation of the weakly compact subsets of the dual of $H^\infty(U)$ and is due to CHAUMAT. It is equivalent to the fact that β_1 is the Mackey topology on

$H^\infty(U)$. Unfortunately, the only existing proofs are too long and technical to be reproduced here.

2.12. Proposition: Let (f_n) be a sequence in $(H^\infty(U),\beta)'$ which converges pointwise to a functional f on H^∞. Then f is β-continuous.

2.13. Proposition: Let K be a bounded subset of the dual $L^1/(H^\infty)^\circ$ of $(H^\infty(U),\beta_1)$. Then the following are equivalent:

1) K is weakly relatively compact;

2) $\lim_{C\to\infty} \sup_{f\in K} (\inf \|f - \pi x\| : x \in CB_{H^\infty}) = 0.$

(π is the projection from $L^1(\partial U)$ onto the dual of $H^\infty(U)$).

2.14. Corollary: $(H^\infty(U),\beta_1)$ is a Mackey space.

Proof: The condition 2) means that for $\varepsilon > 0$ there is a $C > 0$ so that $K \subseteq CB_1 + \varepsilon B_2$ where B_1 is the unit ball of the dual of $H^1(U)$ and B_2 is the unit ball of $L^1(\partial U) / (H^\infty)^\circ$. Hence the result follows from I.1.22.

Note that 2.12 and 2.14 imply that $(H^\infty(U),\beta_1)$ satisfies a closed graph theorem with a separable Fréchet space as range space.

V.3. THE ALGEBRA H^∞

As we have already remarked, $(H^\infty(U),\beta)$ is a topological algebra. In this section, we describe the closed ideals of $H^\infty(U)$. It turns out that these have an especially simple form - in fact, they are all principal ideals. This classification is, of course, of intrinsic interest but it also has a number of useful applications. Following RUBEL and SHIELDS, we use it to give a description of the invariant subspaces of a class of Hardy spaces. In the next section, we give an application of the ideal theory to operators in Hilbert space.

We begin by describing the β-closed maximal ideals of H^∞. Of course, $H^\infty(U)$ is a Saks algebra and can be represented as the projective limit of the spectrum

$$\{A(K) : K \text{ compact in } U\}$$

of Banach algebras where $A(K)$ denotes the set of continuous functions on K which are holomorphic in the interior of K. If $\lambda \in U$,

$$\phi_\lambda : x \longmapsto x(\lambda)$$

is an element of $M_\gamma(H^\infty)$. Thus we have constructed an injection $T : \lambda \longmapsto \phi_\lambda$ from U into $M_\gamma(H^\infty)$.

3.1. <u>Lemma</u>: Suppose that $A \subseteq U$ is not relatively compact in U. Then $\widetilde{A} := \{\phi_\lambda : \lambda \in A\}$ is not β-equicontinuous.

Proof: If $\widetilde{A}$ were equicontinuous, there would be an $r < 1$ and an $\varepsilon > 0$ so that for each $f \in \widetilde{A}$, $|f(x)| < 1/2$ whenever $x \in B_{\| \|}$ with $|x| \leq \varepsilon$ on the set $\{\lambda : |\lambda| \leq r\}$ (I.1.7). Choose $n \in \mathbb{N}$ so that $r^n \leq \varepsilon$. There is a $\lambda_0 \in A$ so that $2|\lambda_0|^n \geq 1$. Then $x : \lambda \longmapsto \lambda^n$ satisfies the above conditions and $|\phi_{\lambda_0}(x)| \geq 1/2$ - contradiction.

3.2. Proposition: T is a homeomorphism from U onto $M_\gamma(H^\infty)$.

Proof: We first show that T is surjective: if $\phi \in M_\gamma(H^\infty)$, let $\lambda := \phi(\mathrm{id}_U)$ where id_U is the identity function on U. Note that $|\lambda| < 1$. For $(\mathrm{id}_U)^n \longrightarrow 0$ in β and so $\lambda^n \longrightarrow 0$. Then $\phi(z_n) = \phi_\lambda(z_n)$ for each $n \in \mathbb{N}$ where z_n is the function $\lambda \longmapsto \lambda^n$. Hence $\phi = \phi_\lambda$ since the linear span of $\{z_n\}$ is β-dense.

Now T is clearly continuous and is a homeomorphism on the compact sets of U. Hence we need only show that every compact set in $M_\gamma(H^\infty)$ is the image of a compact set in U. But this is precisely the content of 3.1.

Using similar techniques, we can characterise the Saks algebra endomorphisms of $H^\infty(U)$.

3.3. Proposition: Let Φ be an algebra homomorphism from $H^\infty(U)$ into itself. Then Φ is β-continuous if and only if it has the form $x \longmapsto x \circ \phi$ where ϕ is a holomorphic function from U into itself.

Proof: The sufficiency is clear.

Necessity: let $\phi := \Phi(id_U)$. Since $(id_U)^n \longrightarrow 0$ in $(H^\infty(U),\beta)$ we have $\phi^n \longrightarrow 0$ and this is only possible if $\phi(U) \subseteq U$. Then Φ coincides with the mapping $x \longmapsto x \circ \phi$ on z_1 (with the notation of the proof of 3.2) and so on $H^\infty(U)$ (since this is the β-closed algebra generated by the unit and z_1).

It is perhaps not inappropriate to give here a classical result of KAKUTANI which can be reinterpreted as stating that an algebra isomorphism from $H^\infty(U)$ onto itself is automatically β-continuous.

3.4. Proposition: Let Φ be an algebra isomorphism from $H^\infty(U)$ onto itself. Then Φ is β-continuous and so there is a conformal mapping ϕ from U onto itself so that $\Phi : x \longmapsto x \circ \phi$.

Proof: Let $x \in H^\infty(U)$, $\lambda \in C$. Then λ is in the closure of the range of x if and only if $x-\lambda$ is not invertible in H^∞ and this is equivalent to the non-invertibility of $\Phi(x)-\lambda$. Hence the closures of the ranges of x and $\Phi(x)$ are the same. In particular, if $\phi := \Phi(id_U)$, then the range of ϕ is an open subset of $\mathbb{C}$ (a non-constant holomorphic function is open!) whose closure is $\bar{U}$ i.e. $\phi(U) \subseteq U$.

Now choose $\lambda_0 \in U$, $x \in H^\infty(U)$. We show that

$$(\Phi x)(\lambda_0) = x(\phi(\lambda_0))$$

and the result will follow. We know that $\phi(\lambda_0) \in U$ and

$(id_U - \phi(\lambda_o))$ is a divisor of $x - x(\phi(\lambda_o))$ in the algebra $H^\infty(U)$. Hence $\phi - \phi(\lambda_o)$ is a divisor of $\Phi(x) - x(\phi(\lambda_o))$. Thus $\Phi(x) - x(\phi(\lambda_o))$ vanishes at λ_o which is the required result.

We now recall some notation (cf. HOFFMAN [21], Ch. 5). A function $x \in H^\infty(U)$ is called an <u>inner function</u> if $|x| = 1$ a.e. on ∂U. An inner function without zeros has the form

$$\lambda \longmapsto \alpha \exp\left[-\int_0^{2\pi} \frac{e^{i\theta} + \lambda}{e^{i\theta} - \lambda} \, d\mu(\theta)\right] \qquad (A)$$

where $\alpha \in \mathbb{C}$ with $|\alpha| = 1$ and μ is a non-negative measure on $[0,2\pi]$ which is singular with respect to Lebesgue measures.

If (λ_k) is a sequence (with "multiplicities") in $U \setminus \{0\}$ so that $\Sigma(1 - |\lambda_k|) < \infty$ then the <u>Blaschke product</u>

$$\prod_k \frac{\overline{\lambda}_k(\lambda_k - \lambda)}{|\lambda_k|(1 - \overline{\lambda}_k\lambda)} \qquad (B)$$

converges for each $\lambda \in U$ and defines an inner function whose zeros are precisely (λ_k) (up to multiplicities). On the other hand, if x is an inner function and (λ_k) is the sequence of zeros ($\neq 0$) of x then $\Sigma(1 - |\lambda_k|) < \infty$ and x can be expressed in the form

$$x(\lambda) = \alpha\lambda^r s(\lambda)B(\lambda)$$

where s has the form (A), B is the Blaschke produkt (B), $r \in \mathbb{N}$ and $\alpha \in C$ with $|\alpha| = 1$.

A function $x \in H^1(U)$ is an <u>outer function</u> if it has the form

$$x : \lambda \longmapsto \alpha \exp\left[\frac{1}{2\pi}\int_0^{2\pi} \frac{e^{i\theta}+\lambda}{e^{i\theta}-\lambda} \ln k(\theta)d\theta\right] \qquad (C)$$

where $k \geq 0$, $\ln k \in L^1([0,2\pi])$ and $|\alpha| = 1$.
Then $k(\theta) = |u(e^{i\theta})|$ a.e. $x \in H^p(U)$ if and only if $k \in L^p$. Every function $x \in H^1(U)$ has a canonical factorisation $x = x_i x_o$ where x_i is inner and x_o is outer. x_o is the outer function defined by the formula (C) with $k : \theta \longmapsto |u(e^{i\theta})|$.

In preparation for our main result on closed ideals, we prove some results on principal ideals generated by inner and outer functions.

3.5. <u>Proposition</u>: If x is an inner function, then xH^∞ is a β-closed ideal of $H^\infty(U)$.

<u>Proof</u>: We show that xH^∞ is β-closed. It suffices to show that if (x_n) is a sequence in H^∞ so that $xx_n \longrightarrow x_o$ for β then x_o is divisible by x. Now $\{xx_n\}$ is norm bounded and hence so is $\{x_n\}$ (since $|x| = 1$ on ∂U). Then (x_n) has a pointwise convergent subsequence (x_{n_k}).
Let $y := \lim x_{n_k}$. Then $xx_{n_k} \longrightarrow xy$ for β and so $x_o = xy \in xH^\infty$.

3.6. <u>Proposition</u>: Let $x \in H^\infty(U)$. Then xH^∞ is β-dense if and only if x is an outer function.

Proof: Sufficiency: let x be an outer function with $\|x\| = 1$. Then $h : \theta \longmapsto \ln|x(e^{i\theta})|$ is integrable and

$$x(\lambda) = \exp\left[\int_0^{2\pi} \frac{\lambda + e^{i\theta}}{\lambda - e^{i\theta}} h(\theta)d\theta\right] \qquad (\lambda \in U)$$

Let $h_n : \theta \longmapsto \min(h(\theta), n)$

$$y_n : \lambda \longmapsto \exp\left[-\int_0^{2\pi} \frac{\lambda + e^{i\theta}}{\lambda - e^{i\theta}} h_n(\theta)d\theta\right]$$

Then

$$xy_n^{-1} : \lambda \longmapsto \exp\left[-\int_0^{2\pi} \frac{\lambda + e^{i\theta}}{\lambda - e^{i\theta}} (h(\theta) - h_n(\theta))d\theta\right]$$

and $(h-h_n)$ is a sequence of non-negative functions which is L^1-convergent to zero. Hence $\|xy_n\| \le 1$ and $xy_n \longrightarrow 1$ pointwise. Thus the constant function 1 lies in the β-closure of xH^∞ and so $\overline{xH^\infty} = H^\infty$.

Necessity: if x has a non-trivial inner factor x_o, we have $xH^\infty \subseteq x_oH^\infty \neq H^\infty$ and so $\overline{xH^\infty} \subseteq \overline{x_oH^\infty} = x_oH^\infty \neq H^\infty$.

We can now prove the main result of this section. We use the following simple Lemma (see HOFFMAN [21], p. 85):

3.7. Lemma: Let F be a non-empty family of inner functions. Then there is a unique inner function x which divides each function of F and is such that each divisor of F is a divisor of x (x is called the greatest common divisor of F).

Proof (sketch): We use the factorisation of an inner function as a product of a Blaschke product and a function of the form (A). The greatest common divisor is the product of the Blaschke product formed from the common zeros of the functions of $\mathcal{F}$ and the function (A) defined by the measure μ which is the supremum of the corresponding measures for the functions of $\mathcal{F}$.

3.8. Proposition: Let J be a non-zero closed ideal of H^∞ and let x_0 be the greatest common divisor of the inner parts of the non-zero functions of J. Then $J = x_0 H^\infty$.

Proof: We have $J \subseteq x_0 H^\infty$ by definition of x_0. By replacing J by $\tilde{J}$, the set of functions of the form x/x_0 $(x \in J)$ we can reduce to the case $x_0 = 1$ i.e. we can assume that the greatest common divisor of the inner parts of the functions of J is 1. We must then show that $J = H^\infty$.

Let $y \in L^1(\partial U)$ be such that $\int_{\partial U} xy = 0$ for each $x \in J$.

We show that $y \in H_0^1 = (H^\infty)^0$ and this will suffice by the bipolar theorem (we are regarding (H^∞, β) as a subspace of $(L^\infty(\partial U), \beta_\sigma))$.

For any $x \in J$, we have

$$\int_{\partial U} \lambda^n x(\lambda) y(\lambda) d\lambda = 0$$

for each $n \in \mathbb{N}$. Hence, by the Riesz Lemma (cf. HOFFMAN [21], p. there is an H^1-function H_x (vanishing at 0) so that

$$xy = H_x$$

on ∂U. Hence y agrees (a.e.) with the non-tangential limit of the meromorphic function H_x/x on ∂U. In particular, H_x/x is independent of x. Hence it is, in fact, analytic (since the x's of J do not have a common zero in U). Thus y is the non-tangential limit of an analytic function (which vanishes at zero since each H_x does) - q.e.d.

We now show that the above result can be used to characterise invariant subspaces of several important spaces of analytic functions on the disc. We consider a topological vector space E whose elements are holomorphic functions on U (with the usual algebraic operations) and are of bounded characteristic i.e. expressible as a quotient x/y of two functions x,y in $H^\infty(U)$. In addition, we suppose that

1) if E_1 is a closed subspace of E then $E_1 \cap H^\infty$ is β-closed in $H^\infty(U)$;

2) if $x \in H^\infty$, the mapping

$$x \longmapsto xy$$

maps E continuously into itself;

3) if $x \in E$ has a representation $\frac{y_1 z}{y_2}$ where z is outer and y_1 and y_2 are inner functions without a non-trivial common inner factor, then $z/y_2 \in E$;

4) $H^\infty(U)$ is dense in E.

The Hardy spaces H^p $(1 \le p < \infty)$ satisfy these conditions (cf. RUBEL and SHIELDS [31], § 5.9).

3.9. Proposition: Let E satisfy the above conditions and let E_1 be a closed subspace of E which is invariant (i.e. $xE_1 \subseteq E_1$ for each $x \in H^\infty$). Then there is an inner function x_o so that $E_1 = x_o E$.

Proof: We first note that exactly as in the proof of 3.6, one can (using condition 1) above) show that if x_o is an outer function in E, then the constant function 1 lies in the E-closure of $x_o H^\infty$.

Now let x be an element of E_1 - x has a representation of the form $y_1 z/y_2$ where z is outer and y_1, y_2 are inner functions without common factor. Then $y_1 \in E_1$. For there is a sequence (x_n) of bounded analytic functions so that $zx_n \longrightarrow 1$ in E. Then, by 2), $y_1 z x_n = x y_2 x_n \longrightarrow y_1$ in E and so $y_1 \in E_1$.

Let $\tilde{E}_1 = E_1 \cap H^\infty$ - then $\tilde{E}_1$ is a β-closed ideal in $H^\infty(U)$ (by 1)) There is an inner function x_o so that $\tilde{E}_1 = x_o H^\infty$. We claim that $E_1 = x_o E$.

a) $E_1 \subseteq x_o E$: choose $x \in E_1$ and let $x = \dfrac{y_1 z}{y_2}$ where z is outer and y_1, y_2 are inner functions without a common factor. Then $y_1 \in E_1$ (see above) and so $y_1 \in \tilde{E}_1$. Thus there is an inner function y_3 so that $y_1 = x_o y_3$. Then $x = \dfrac{x_o y_3 z}{y_2}$. Now $z/y_2 \in E$ (by 3)) and so $\dfrac{y_3 z}{y_2} \in E$. Hence $x \in x_o E$.

b) $x_o E \subseteq E_1$: if $x \in E$, there is a net $(x_\alpha)_{\alpha \in A}$ in $H^\infty(U)$ which converges (in E) to x. Then $x_o x_\alpha \longrightarrow x_o x$ and so $x_o x \in E_1$ since $x_o x_\alpha \in E_1$.

3.10. Corollary: The non-zero closed invariant subspace of H^p $(1 \le p < \infty)$ are precisely those of the form $x_o H^p$ (x_o an inner function).

3.11. Remark: Note that 3.9 does not state that a subspace of the form $x_o E$ is closed in E. However, it is easy to see that this is true in the case of the Hardy spaces H^p.

V.4. THE H^∞-FUNCTIONAL CALCULUS FOR COMPLETELY NON UNITARY CONTRACTIONS

If T is a contraction in a Banach space (i.e. $\|T\| \le 1$) then the classical functional calculus can be developed for functions analytic on a neighbourhood of the closed unit disc. SZ. NAGY and FOIAŞ have shown how this can be improved to give a functional calculus for functions in $H^\infty(U)$ for certain contractions in Hilbert space. In this section we give a brief description of some of their results from the point of view of the strict topology. This allows a simpler and more direct approach. We begin with some preliminary results and definitions. A detailed account can be found in Chapter I of [25].

4.1. Definition: A contraction T in a Hilbert space H is completely non unitary if there is no non-trivial subspace M of H which is invariant under T and T^* and is such that $T|_M$ is unitary. We write c.n.u. contraction for a completely non unitary contraction.

For every contraction T there is a unique orthogonal decomposition $H = H_0 \oplus H_1$ of H so that H_0 and H_1 reduce T (i.e. are invariant under T and T^*) and are such that $T|_{H_0}$ is unitary and $T|_{H_1}$ is c.n.u. (see [25], Th. I.3.2). H_0 is characterised as the space

$$\{x \in H : \|T^n x\| = \|x\| = \|T^{*n} x\| \quad \text{for each } n \in \mathbb{N}\}$$

T is c.n.u. if and only if $H_0 = \{0\}$.

Our construction uses the following important result:

4.2. Proposition ([25], Th. I.4.1): Let T be a contraction in H. Then there is a Hilbert space K which contains H as a subspace and a unitary operator U on K so that

1) $K = \bigvee_{n=-\infty}^{\infty} U^n H$, the closed subspace of K generated by $\bigcup_{n=-\infty}^{\infty} U^n H$;

2) $T^n x = PU^n x$ ($x \in H$, $n \in \mathbb{N}$) where P is the orthogonal projection from K onto H.

With the notation of 4.2, we introduce the subspaces

$$L := \overline{(U-T)H}; \qquad L^* := \overline{(U^*-T^*)H}$$

of K. M(L) (resp. M(L*)) denotes the closed subspace of K spanned by $M_0(L) := \bigcup_{n=-\infty}^{\infty} U^n L$ (resp. $M_0(L^*) := \bigcup_{n=-\infty}^{\infty} U^n L^*$).

4.3. <u>Proposition</u> ([25], Th. II.1.1): L and L* are wandering subspaces for U, that is

$$U^m L \perp U^n L \quad \text{and} \quad U^m L^* \perp U^n L^* \qquad (m,n \in \mathbb{N},\ m \neq n).$$

4.4. <u>Proposition</u> ([25], Prop. II.1.4): Let T be a contraction. Then

$$H_0 = (M(L) \vee M(L^*))^\perp$$

(where H_0 is as in 4.1 and $M(L) \vee M(L^*)$ denotes the closed linear subspace spanned by $M(L) \cup M(L^*)$).
Hence if T is c.n.u. then $M(L) \vee M(L^*) = K$.

We now denote by Pol(U) the space of functions $x \in H^\infty(U)$ which are the restrictions of polynomials to U. Then, of course, Pol(U) is β-dense in $H^\infty(U)$, in fact $\mathrm{Pol}(U) \cap B_{H^\infty}$ is τ_K-dense in B_{H^∞} (2.2). Thus if we regard Pol(U) as a Saks space with the structures $(\|\ \|, \tau_K)$, $H^\infty(U)$ is its completion in the sense of I.3.6. Now if $p \in \mathrm{Pol}(U)$, and T is a contraction on H, we denote by p(T) the operator obtained by formal substitution of T in p. Then

$$\Phi : p \longmapsto p(T)$$

is an algebra homomorphism from Pol(U) into L(H). In the following, we shall show that if T is completely non-unitary, then

Φ is β-β_σ-continuous (resp. β_1-β_s-continuous) and so can be extended to a continuous algebra homomorphism from H^∞ into $L(H)$.

4.5. Lemma (von Neumann's inequality): If $p \in Pol(U)$ and T is a contraction then $\|p(T)\| \leq \|p\|$. In other words Φ is norm-bounded (and, in fact, $\|\Phi\| \leq 1$).

Proof: Let U be the unitary operator of 4.2. Then of course $\|p(U)\| \leq \|p\|$. On the other hand, $p(T)x = Pp(U)x$ $(x \in H)$ (by 4.2.2)) and so $\|p(T)\| \leq \|p(U)\|$.

If T is a contraction, U as in 4.2, then for each $x,y \in H$ the function

$$\lambda \longmapsto ((\lambda - U)^{-1} x \mid y)$$

is analytic on $\complement\overline{U}$.

4.6. Lemma: Suppose that $x,y \in M_0(L)$ (resp. $M_0(L^*)$). Then

1) the function $\lambda \longmapsto ((\lambda - U)^{-1} x \mid y)$ has an analytic extension to $\mathbb{C} \setminus \{0\}$;

2) the following equality holds

$$\|p(U)x\| = \|p\|_2 \, \|x\|.$$

Proof: 1) We use the expansions

$$\sum_{n=0}^{\infty} \lambda^{-1-n} (U^n x \mid y) = \sum_{n=0}^{\infty} \lambda^{-1-n} (x \mid U^{*n} y)$$

which have only a finite number of non-zero terms since L (resp. L^*) is a wandering subspace for U.

2) If p is the polynomial $\lambda \longmapsto \sum_{k=0}^{n} a_k \lambda^k$, we have

$$\|p(U)x\|^2 = (p(U)x|p(U)x) = (\sum_k a_k U^k x \mid \sum_l a_l U^l x)$$
$$= \sum_{k,l} a_k \bar{a}_l (U^k x|U^l x)$$
$$= (\sum_k |a_k|^2)\|x\|^2.$$
$$= \|p\|_2^2 \|x\|^2.$$

4.7. Lemma: Let Φ be a norm bounded linear mapping from Pol(U) into L(H). Suppose that M_1 and M_2 are subspaces of H which are invariant under $\Phi(p)$ for each $p \in$ Pol(U) and are such that $M_1 \cup M_2$ is total in H. Then

1) if for each $x,y \in M_1$ (resp. M_2) the mapping

$$p \longmapsto (\Phi(p)x|y)$$

is β-continuous, Φ is β-β_σ-continuous;

2) if for each $x \in M_1$ (resp. M_2) the mapping

$$p \longmapsto \Phi(p)x$$

is β_1-continuous, Φ is β_1-β_s-continuous.

Proof: 1) The hypotheses imply that Φ is β-β_σ-continuous when regarded as a mapping with values in $L(M_1)$ (resp. $L(M_2)$). Let P_{M_1} denote the orthogonal projection onto M_1. Then if $x \in M_1$, $y \in M_2$, $p \in$ Pol(U), it follows that $(\Phi(p)x|y) = (\Phi(p)x|Py)$ and so

$$p \longmapsto (\Phi(p)x|y)$$

is β-continuous. The case $x \varepsilon M_2$, $y \varepsilon M_1$ is handled similarly and so Φ is β-β_σ-continuous on the unit ball of Pol(U) since the weak topology y induced by $M_1 \cup M_2$ coincides with the weak topology on $B_{L(H)}$.

The proof of 2) is similar.

4.8. <u>Lemma</u>: Let T be a c.n.u. contraction. Then the mapping $\Phi : p \longmapsto p(T)$ from Pol(U) into L(H) is β-β_σ and β_1-β_s-continuous.

<u>Proof</u>: Φ is norm-bounded by 4.5.

β-β_σ continuity: let (p_n) be a sequence in Pol(U) so that $p_n \longrightarrow 0$ in β. Since $(p_n(T)x|y) = (p_n(U)x|y)$ for each $x,y \varepsilon H$ it will be sufficient to show that $p_n(U) \longrightarrow 0$ in $(L(K),\beta_\sigma)$. Applying 4.7 with $M_1 = M(L)$, $M_2 = M(L^*)$, we need only show that if $x,y \varepsilon M_0(L)$ (resp. $M_0(L^*)$) then $(p_n(U)x|y) \longrightarrow 0$. But then

$$2\pi i(p_n(U)x|y) = \int_{|\lambda|=R>1} p_n(\lambda)((\lambda-U)^{-1}x|y)d\lambda$$
$$= \int_{|\lambda|=r<1} p_n(\lambda)((\lambda-U)^{-1}x|y)d\lambda$$

(by 4.6.1)) and the latter converges to zero since $p_n \longrightarrow 0$ uniformly on $\{\lambda : |\lambda| = r\}$.

The β_1-β_s-continuity is proved similarly using 4.6.2).

4.9. Proposition: Let T be a c.n.u. contraction on H. Then there is an algebra homomorphism $\Phi : H^\infty(U) \longrightarrow L(H)$ so that

1) $\Phi(p) = p(T)$ for $p \in Pol(U)$;

2) Φ is $\| \|-\| \|$, $\beta-\beta_\sigma$ and $\beta_1-\beta_s$-continuous;

3) if T is normal, $\Phi(x) = x(T)$ in the sense of the functional calculus for normal operators;

4) $\Phi(x) = P \circ x(U)$ on H.

Proof: By 4.8 and the remarks before 4.5, we can extend the operator $\Phi : Pol(U) \longrightarrow L(H)$ to a $\beta-\beta_\sigma$-continuous linear operator from $H^\infty(U)$ into L(H). The properties 2) - 4) can be deduced easily.

4.10. Remark: If x is an element of $H^\infty(U)$, it is natural to denote $\Phi(x)$ by x(T). Property 4.9.2) provides two natural methods of calculating x(T). If $0 < r < 1$, denote by x_r the function $\lambda \longrightarrow x(r\lambda)$. Then x_r is analytic on a neighbourhood of $\bar{U}$ and so $x_r(T)$ can be defined by the classical functional calculus. $x_r \longrightarrow x$ in $(H^\infty(U),\beta_1)$ and so $x(T) = s - \lim_{r\to 1-} x_r(T)$ (the strong limit). This was used as the definition of x(T) by SZ.-NAGY and FOIAŞ. Alternatively, if $x \in H^\infty$ has the Taylor expansion $\sum_{k=0}^{n} a_k\lambda^k$ then

$$x(T) = s - \lim_{n\to\infty} \frac{1}{n+1} (s_0 + \dots + s_n)$$

where $s_n = \sum_{k=0}^{n} a_k T^k$.

A c.n.u. contraction T ($\neq 0$) on H is <u>of class</u> C_0 if there is a non-zero $x \in H^\infty(U)$ so that $x(T) = 0$. Let

$$J := \{x \in H^\infty(U) : x(T) = 0\}.$$

Then J is a non-zero, β-closed, proper ideal of H^∞. By 3.8 it has the form $x_0 H$ where x_0 is an inner function. x_0 satisfies the following properties:

1) $x_0(T) = 0$;

2) if $y \in H^\infty$ is such that $y(T) = 0$ then x_0 is a divisor of y.

x_0 is called the <u>minimal function</u> of T and denoted by m_T. It plays a role corresponding to that of the minimal polynomial of a matrix (i.e. finite dimensional linear operator) and contains useful information on T. As examples we give two results which follow easily from the theory developed here.

If x is an inner function with factorisation of the form

$$x(\lambda) = \lambda^r s(\lambda) B(\lambda)$$

(cf. the remarks before 3.5) we define the <u>generalised zero set</u> $Z(x)$ to be the union of the following three sets:

1) the zeros of the Blaschke product (together with 0 if $r \geq 1$);

2) the accumulation points of the sets 1);

3) the (closed) support of the singular measure in the canonical representation of s (see equation (A) after Prop. 3.4).

(For some remarks on the functiontheoretical significance of this set see HOFFMAN [21], pp. 68,69).

4.11. Proposition: If T is a c.n.u. contraction of class C_o with minimal function m_T, then the spectrum $\sigma(T)$ of T is given by the equation

$$\sigma(T) = Z(m_T).$$

Proof: (i) $\sigma(T) \subseteq Z(m_T)$: suppose that $\lambda_o \in \overline{U} \setminus Z(m_T)$. Then

$$x : \lambda \longmapsto \frac{1}{\lambda-\lambda_o}\left[m_T(\lambda_o) - m_T(\lambda)\right]$$

is in H^{∞} and so, "substituting T", we have

$$\begin{aligned}(\lambda_o I - T)x(T) &= x(T)(\lambda_o I - T)\\ &= m_T(\lambda_o)I - m_T(T)\\ &= m_T(\lambda_o)I.\end{aligned}$$

Hence $\lambda_o \notin \sigma(T)$.

(ii) $Z(m_T) \cap U \subseteq \sigma(T)$: let $\lambda_o \in U \cap Z(x)$. Then we can write

$$m_T(\lambda) = \frac{\lambda-\lambda_o}{1-\lambda_o\lambda}\, x_i(\lambda)$$

where x_i is inner. Substituting T once again gives

$$0 = (T - \lambda_o I)x_i(T)$$

and so $(T - \lambda_o I)$ is not injective since $x_i(T) \neq 0$ (m_T is not a divisor of x_i !).

(iii) $Z(m_T) \cap \partial U \subseteq \sigma(T)$: suppose that λ_o is in $\rho(T) \cap \partial U$ ($\rho(T) := \mathbb{C} \setminus \sigma(T)$ is the resolvent of T). Then by part (ii) of the present proof, λ_o is not an accumulation of the zeros of the Blaschke product of m_T (since $\rho(T)$ is open). The proof will then be completed if we show that for any open arc W in $\rho(T) \cap \partial U$, $\mu(W) = 0$ where is the singular measure involved in the canonical representation of the singular part of m_T. If this is not the case, there is a closed subarc K of W with $\mu(K) > 0$. Then the inner function

$$x_1 : \lambda \longmapsto \exp\left[-\int_K \frac{e^{i\theta}+\lambda}{e^{i\theta}-\lambda}\, d\mu(\theta)\right]$$

is a non-constant divisor of m_T and so is $x_2 := m_T/x_1$, x_2 is inner and

$$x_1(T)x_2(T) = m_T(T) = 0.$$

$x_1(T)$ is thus not injective (for otherwise we would have $x_2(T) =$ which would imply that m_T is a divisor of x_2). Hence the subspace

$$H_1 := \{y \in H : x_1(T)y = 0\}$$

is a non-zero invariant subspace for T. Let $T_1 := T|_{H_1}$. Then m_{T_1}, the minimal function of T_1, is a divisor of m_T and so has the form

$$\lambda \longmapsto \exp\left[-\int_0^{2\pi} \frac{e^{i\theta}+\lambda}{e^{i\theta}-\lambda}\, d\mu_1(\theta)\right]$$

where μ_1 is a non-negative singular measure which is bounded above by μ on K and vanishes outside of K. Thus $Z(m_{T_1}) \subseteq K$

and so by part (i) of the present proof,

$$\sigma(T_1) \subseteq Z(m_{T_1}) \subseteq K.$$

However, it is easy to see that $K \subseteq \rho(T_1)$ and so

$$\sigma(T_1) \subseteq \rho(T_1)$$

a contradiction. Hence $\mu(W) = 0$ and the proof is finished.

4.12. Proposition: If T is a c.n.u. contraction of class C_0 and x is an inner divisor of m_T then

$$H_0 := \{y \in H : x(T)y = 0\}$$

is invariant for T.

As a Corollary, we obtain the following partial result on the famous invariant subspace problem.

4.13. Corollary: If $\dim(H) > 1$ then every c.n.u. contraction of class C_0 on H has a non-trivial invariant subspace.

The proof of 4.12 is obvious (in fact, one can show that H_0 is hyperinvariant, that is invariant for every operator which commutes with T). The Corollary follows immediately if m_T has a non-trivial factorisation. Otherwise, it has the form

$$\lambda \longmapsto \alpha \frac{\lambda-\lambda_0}{1-\lambda_0\lambda}$$

for some $|\alpha| = 1$, $\lambda_0 \in U$. An easy calculation shows then that $T = \lambda_0 I$ and so has non-trivial invariant subspaces.

V.5. NOTES

As mentioned in the introduction to this chapter, the topology β on H^∞ was first introduced by BUCK [10]. It was systematically studied by RUBEL and SHIELDS [31] and RUBEL and RYFF [30]. The results of section 1 can be found in these papers (usually with different proofs).

The results and methods of § 2 are mostly new - in particular, the topology β_1 seems to be new. The fact that β is not the Mackey topology is due to CONWAY [13]. This result has been generalised to more general domains by RUBEL and RYFF. BIERSTEDT [6] and [7] has obtained tensor product representations for $H^\infty(U \times U)$ and considers the approximation property using different methods. BIRTEL [8] and BIRTEL, DUBINSKY [9] have considered Banach space tensor products of H^∞-spaces. MOONEY's theorem (2.12) seems to have been conjectured originally by A.E. TAYLOR (see PIRANIAN, SHIELDS, WELLS [26]). For proofs see AMAR [1] and MOONEY [24]. BARBEY [2] and [3] has given a version of this theorem in the setting of spaces of abstract analytic functions. CHAUMAT's theorem is in [12].
The topology β_1 can be defined for domains G more general than U. It is necessary that the boundary of G be rectifiable and regular enough that H^∞-functions have (non-tangential) boundary values and be recoverable from then. A nice discussion of this problem can be found in ZALCMAN [37].

§ 3 is based on the work of RUBEL and SHIELDS. Concerning the main result (3.9) (which is originally due to SRINAVASAN), they write: "we prove the result here by using Beurling's characterisation of the closed invariant subspaces of H^2.... It would be good to find a direct intrinsic proof".
The proof given here is adapted from the proof of the corresponding result for the space of functions in H^∞ which have continuous extension to $\overline{U}$ (a result due to BEURLING and RUDIN - see HOFFMAN [21], Ch. 6). The theorem (3.4) of KAKUTANI was published in [22]. We have taken the proof from HOFFMAN [21], p. 144.

U, the γ-spectrum of $H^\infty(U)$, is naturally embedded in the spectrum of a Banach slgebra $(H^\infty, \| \ \|)$. The famous problem whether U is a dense subspace (the "Corona problem") was solved positively by CARLESON in [11].

In § 4 we have given a brief introduction to part of the theory of completely non-unitary contractions in Hilbert space. The theory is developed in detail by SZ.-NAGY and FOIAŞ in [25]. We have used some ideas of TON-THAT-LONG [36].

Further references on (H^∞, β) are BARTELT ([4] and [5]), and SHAPIRO ([33], [34] and [35]).

There are three recent survey articles on spaces of bounded analytic functions - GAMELIN [17], RUBEL [29] and SARASON [32].

The paper of ZALCMAN [37] also contains an excellent account of current work on algebras of analytic functions.

In §§ 2 and 3 we have considered only functions on the open unit disc in order to simplify the presentation. However, many of the results can be extended to more general regions by simple methods. For example, 2.11 holds when U is replaced by a region which is the finite union of disjoint, simply connected domains (as BIERSTEDT has shown). The question of whether these results hold for general regions G can lead to complicated analytic questions. For example, RUBEL and RYFF give a fairly detailed account of extensions of Proposition 3.2. RUBEL and SHIELDS have introduced the notion of inner and outer functions for general regions (even in higher dimensions) but little seems to be known about them.

REFERENCES FOR CHAPTER V.

[1] E. AMAR Sur un théoreme de Mooney relatif aux fonctions analytiques bornées, Pac. J. Math. 49 (1973) 311-314.

[2] K. BARBEY Ein Satz über abstrakte analytische Funktionen, Arch. der Math. 26 (1975) 521-527.

[3] Zum Satz von Mooney für abstrakte analytische Funktionen (Preprint).

[4] M. BARTELT Multipliers and operator algebras on bounded analytic functions, Pac. J. Math. 37 (1971) 575-584.

[5] Approximation in operator algebras on bounded analytic functions, Trans. Amer. Math. Soc. 170 (1972) 71-83.

[6] K.D. BIERSTEDT Injektive Tensorprodukte und Slice-Produkte gewichteter Räume stetiger Funktionen, Jour. reine angew. Math. 266 (1974) 121-131.

[7] Gewichtete Räume stetiger vektorwertiger Funktionen und das injektive Tensorprodukt II, Jour. f. reine angew. Math. 260 (1973) 133-146.

[8] F. BIRTEL Slice algebras of bounded analytic functions, Math. Scand. 21 (1967) 54-60.

[9] F. BIRTEL, E. DUBINSKY Bounded analytic functions in two complex variables, Math. Z. 93 (1966) 299-310.

[10] R.C. BUCK Algebraic properties of classes of analytic functions, Seminars on analytic functions II, 175-188 (Princeton, 1957).

[11] L. CARLESON Interpolation by bounded analytic functions and the corona problem, Ann. of Math. 76 (1962) 542-559.

[12] J. CHAUMAT Une généralisation d'un théorème de Dunford-Pettis (Sem. Analyse Harmonique d'Orsay No. 85 - 1974).

[13] J.B. CONWAY Subspaces of $(C(S),\beta)$, the space (ℓ^{∞},β) and (H^{∞},β), Bull. Amer. Math. Soc. 72 (1966) 79-81.

[14] J.B. COOPER The H^{∞}-functional calculus for completely non-unitary contractions, Math. Balkanica 4.15 (1975) 89-90.

[15] F. DELBAEN Weakly compact sets in H^1, Pac. J. Math. 63 (1976) 367-369.

[16] T.W. GAMELIN Uniform algebras (New Jersey, 1969).

[17] The algebra of bounded analytic functions Bull. Amer. Math. Soc. 79 (1973) 1095-1108.

[18] A. GROTHENDIECK Sur certains espaces de fonctions holomorphes, I,II, J. reine angew. Math. 192 (1953) 35-64, 77-95.

[19] V.P. HAVIN Weak completeness of the space L^1/H_0^1 (Russian), Vestnik Leningrad Univ. 13 (1973) 77-81.

[20] H. HELSON Lectures on invariant subspaces (New York, 1964).

[21] K. HOFFMAN Banach spaces of analytic functions (New Jersey, 1962).

[22] S. KAKUTANI Rings of analytic functions, Lectures on functions of a complex variable, 71-83 (AnnArbor, 1955).

[23] A. KERR-LAWSON A filter description of homomorphisms of H^∞, Can. J. Math. 17 (1965) 734-757.

[24] M. MOONEY A theorem on bounded analytic functions, Pac. J. Math. 43 (1972) 457-463.

[25] B. SZ.-Nagy, C. FOIAŞ Harmonic analysis of operators on Hilbert space (Amsterdam, 1970).

[26] G. PIRANIAN, A.L. SHIELDS, J.H. WELLS Bounded analytic functions and absolutely continuous measures, Proc. Amer. Math. Soc. 18 (1967) 818-826.

[27] M. von RENTELN Finitely generated ideals in the Banach algebra H^∞, Collect. Math. 26 (1975) 115-125.

[28] J.R. ROSAY Une equivalence du Corona problème et un problème d'ideals dans $H^\infty(D)$, Jour. Func. Anal. 7 (1971) 71-84.

[29] L.A. RUBEL Bounded convergence of analytic functions, Bull. Amer. Math. Soc. 77 (1971) 13-24.

[30] L.A. RUBEL, J.V. RYFF The bounded weak-star topology and the bounded analytic functions, J. Functional Anal. 5 (1970) 167-183.

[31] L.A. RUBEL, A.L. SHIELDS The space of bounded analytic functions in a region, Ann. Inst. Fourier 16 (1966) 235-277.

[32] D. SARASON Algebras of functions on the unit circles, Bull. Amer. Math. Soc. 79 (1973) 286-299.

[33] J.H. SHAPIRO Weak topologies on subspaces of C(S), Trans. Amer. Math. Soc. 157 (1971) 471-479.

[34] J.H. SHAPIRO The bounded weak star topology and the general strict topology, J. Func. Anal. 8 (1971) 275-286.

[35] Noncoincidence of the strict and strong operator topologies, Proc. Amer. Math. Soc. 35 (1972) 81-81.

[36] TON-THAT-LONG Sur le calcul fonctionnel d'une contraction complètement non unitaire, Proc. Func. Anal. Week, 26-36 (Aarhus, 1969).

[37] L. ZALCMAN Bounded analytic functions on domains of infinite connectivity, Trans. Amer. Math. Soc. 144 (1969) 241-269.

APPENDIX

Introduction: In this appendix we assemble some results on Saks spaces from a category-theoretical point of view. It is our opinion that this approach puts a number of the results of the previous Chapters in their proper perspective. The *leitmotiv* is the establishment of a duality theory for various important categories.

In the first section, we define formally the category of Saks spaces, list some of its simple property and identify its dual category. We then use these categories to establish in § 2 a duality theory for uniform spaces and for compactologies. The latter are a generalisation of the compact spaces and allow us to give a symmetrical form of the duality theory for completely regular spaces established in Chapter II. In § 3 we consider some important functors and bifunctors on categories of Saks spaces and their duals.

In the fourth section, we use the duality theory of § 2 to give a duality theory for certain classes of semigroups and groups.

In the last section we describe an extension process for categories which formalises the transition from Banach spaces to Saks spaces and demonstrate thereby that this is a special exapmle of a construction which includes, for example, the generalisation from Banach spaces to locally convex spaces or to convex bornological spaces.

A.1. CATEGORIES OF SAKS SPACES

1.1. The categories PSS, SS and CSS: We introduce three categories PSS, SS and CSS with the following objects:

PSS - triples $(E,\| \|,\tau)$ where $(E,\| \|)$ is a normed space, τ is a locally convex topology on E and $B_{\| \|}$, the unit ball of $(E,\| \|)$ is τ-bounded;

SS - triples $(E,\| \|,\tau)$ as above with the additional assumption that $B_{\| \|}$ is τ-closed and $\tau = \gamma[\| \|,\tau]$;

CSS - as for SS with the additional assumption that (E,τ) be complete.

In each case, if $(E,\| \|,\tau)$, $(F,\| \|_1,\tau_1)$ are objects of one of the above categories, the morphisms from E into F are the linear contractions $T : E \longrightarrow F$ so that $T|_{B_{\| \|}}$ is τ-τ_1-continuous.

1.2. Proposition: SS and CSS are full, reflective subcategories of PSS.

Proof: The reflection from PSS into SS is obtained by first employing the procedure described after I.3.2 (i.e. replacing $B_{\| \|}$ by its τ-closure) and then replacing τ by $\gamma[\| \|,\tau]$. The reflection from PSS into CSS is obtained by composing this functor with the completion functor (I.3.6).

We remark that all of the above categories are complete and cocomplete. Products and equalisers in PSS are formed as described in I.3.5 and I.3.7. The construction of sums and co-equalisers is a little more complicated and we shall not describe it explicitely here. The various limits in SS and CSS are constructed by first forming them in PSS and then reflecting down into SS and CSS respectively.

1.3. <u>The categories PMW and MW</u>: An object of PMW is a triple $(E, \| \|, \mathcal{B})$ where $(E, \| \|)$ is a normed space, $\mathcal{B}$ is a bornology on E so that each $B \in \mathcal{B}$ is $\| \|$- bounded and, in addition, if $B \in \mathcal{B}$, there is a locally convex topology τ_B on the corresponding normed space E_B so that $(E_B, \| \|_B, {}_B)$ is a Saks space with B τ_B-compact. If $(E, \| \|, \mathcal{B})$, $(E_1, \| \|_1, \mathcal{B}_1)$ are objects of PMW, a <u>PMW-morphism</u> from E into E_1 is a linear norm-contraction T from E into E_1 which satisfies the following condition:

For each $B \in \mathcal{B}$, there is a $B_1 \in \mathcal{B}_1$ so that $T(B) \subseteq B_1$ and $T|_{E_B}$ is a Saks space morphism from E_B into $(E_1)_{B_1}$.

If $(E, \| \|, \mathcal{B})$ is an object of PMW, we put

$$\widetilde{\mathcal{B}} := \{C \subseteq E : \text{for each } \varepsilon > 0 \text{ there is a } B \in \mathcal{B} \text{ so that } C \subseteq B + \varepsilon B_{\| \|}\}$$

MW denotes the full subcategory of PMW whose objects $(E, \| \|, \mathcal{B})$ satisfy the conditions $\mathcal{B} = \widetilde{\mathcal{B}}$ and $(E, \| \|)$ is a Banach space. In addition, we require the existence of a (separated) locally convex topology on E which coincides with τ_B on each B. An object of MW is called a <u>CoSaks space</u>.

1.4. <u>Examples</u>: I. If $(E,\|\ \|,\tau)$ is a Saks space, then its dual space E'_γ has a natural PMW-structure $(\|\ \|,\mathcal{B}_\gamma)$ where $\|\ \|$ is the dual norm and $\mathcal{B}_\gamma$ is the family of absolutely convex, weakly closed γ-equicontinuous subsets of E'_γ. If $B \in \mathcal{B}_\gamma$ then τ_B is defined to be the weak topology $\sigma(E'_\gamma,E)$.
Then $(E'_\gamma,\|\ \|,\mathcal{B}_\gamma)$ is even an object of MW by I.1.22. In particular, if S is a completely regular space, then the spaces $M_t(S)$, $M_\tau(S)$ and $M_\sigma(S)$ of measures on S, are CoSaks spaces.

II. Let S be a completely regular space. $C^\infty(S)$ has a natural MW-structure $(\|\ \|,\mathcal{B}_{equ})$ where $\|\ \|$ is the supremum norm and $\mathcal{B}_{equ}$ is the set of uniformly bounded, pointwise closed equicontinuous subsets of $C^\infty(S)$. For $B \in \mathcal{B}_{equ}$, the topology τ_B is that of pointwise convergence on S.

III. We can generalise II as follows: let S be a uniform space and denote by $U^\infty(S)$ the vector space of bounded, uniformly continuous functions from S into $\mathbb{C}$. Then $U^\infty(S)$ has a natural CoSaks space structure defined as above (replacing "equicontinuous" by "uniformly equicontinuous"). If S is a completely regular space and we regard it as a uniform space with the <u>fine uniformity</u> (i.e. the finest uniform structure compatible with its topology), then the CoSaks structures defined in this section and in II coincide.

IV. If $(E, \| \|)$ is a Banch space, we can regard E as a CoSaks space by defining $\mathcal{B}$ to be the family of all absolutely convex, compact subsets of $(E, \| \|)$. This allows us to regard the category of Banach spaces as a full subcategory of MW.

1.5. <u>The dual of a CoSaks space</u>: If $(E, \| \|, \mathcal{B})$ is an object of PMW, we define its dual E' to be the space of all linear functionals $f : E \longrightarrow \mathbb{C}$ which are norm-bounded and whose restrictions to each $B \in \mathcal{B}$ are τ_B-continuous. Then E' has a natural Saks space structure $(\| \|, \tau_{\mathcal{B}})$ where $\| \|$ is the dual norm and $\tau_{\mathcal{B}}$ is the topology of uniform convergence on the sets of $\mathcal{B}$.

1.6. <u>Lemma</u>: Let f be a linear functional on the CoSaks space $(E, \| \|, \mathcal{B})$. Then f is norm-bounded (and so an element of E') provided $f|_B$ is τ_B-continuous for each $B \in \mathcal{B}$.

<u>Proof</u>: If f were not bounded on $B_{\| \|}$, we could find a sequence (x_n) in $B_{\| \|}$ so that $|f(x_n)| \geq n^2$. Now $\{x_n/n\}$ is in $\tilde{\mathcal{B}}$ and so in $\mathcal{B}$ and hence f is bounded on this set - contradiction.

1.7. <u>Proposition</u>: If $(E, \| \|, \mathcal{B})$ is a CoSaks space, then $(E', \| \|, \tau_{\mathcal{B}})$ is an object of CSS.

1.8. <u>The duality functors $\mathcal{D}$ and D</u>: The correspondences

$$(E, \| \|, \tau) \longmapsto (E'_\gamma, \| \|, \mathcal{B}_{equ})$$

and

$$(E, \| \|, \mathcal{B}) \longmapsto (E', \| \|, \tau_{\mathcal{B}})$$

are obviously functorial (if we map morphisms into their adjoints) and so we have defined duality functors $\mathcal{D}$ and D where

$$\mathcal{D} : \mathrm{CSS} \longrightarrow \mathrm{MW} \quad \text{and} \quad \mathrm{D} : \mathrm{MW} \longrightarrow \mathrm{CSS}.$$

1.9. Proposition: CSS and MW are quasidual under the functors $\mathcal{D}$ and D.

Proof: The proof that D and $\mathcal{D}$ are mutually adjoint on the left is standard. It then suffices to show that the natural transformation from $D \circ \mathcal{D}$ to the identity functor is equivalent to the identity or, speaking loosely, that $D\mathcal{D}E = E$ for each complete Saks space. This follows from Grothendieck's completeness theorem.

A.2. DUALITY FOR COMPACTOLOGICAL AND UNIFORM SPACES

In § II.2 we established a form of Gelfand-Naimark duality for completely regular spaces. Because the space $(C^\infty(S), \beta_K)$ is not, in general, complete, this duality is not perfect and in this section we show how a complete duality can be obtained by replacing the category of completely regular spaces by that of compactological spaces. We also develop a duality theory for uniform spaces using CoSaks algebras.

2.1. Definition: A compactology on a set S is a family $\mathcal{K}$ of subsets of S, together with a compact (Hausdorff) topology τ_K on each $K \in \mathcal{K}$ so that

1) $\mathcal{K}$ is closed under finite unions and covers S;

2) if $K_1 \subseteq K$ where $K \in \mathcal{K}$ then $K_1 \in \mathcal{K}$ if and only if K_1 is τ_K-closed. If this is the case $\tau_{K_1} = \tau_{K|K_1}$.

A compactological space is a pair $(S,\mathcal{K})$ where $\mathcal{K}$ is a compactology on S. The class of compactological spaces becomes a category when we define morphisms from $(S,\mathcal{K})$ into $(S_1,\mathcal{K}_1)$ to be mappings $f : S \longrightarrow S_1$ so that for each $K \in \mathcal{K}$ there is a $K_1 \in \mathcal{K}_1$ so that $f(K) \subseteq K_1$ and $f_{|K}$ is τ_K-τ_{K_1}-continuous.

If S is a Hausdorff topological space, S has a natural compactology $\mathcal{K}(S)$ (define τ_K to be the induced topology). This construction defines a forgetful functor from the category of Hausdorff topological spaces into the category of compactological spaces.

If $(S,\mathcal{K})$ is a compactological space, we denote by $C^\infty(S)$ the space of bounded morphisms from S into $\mathbb{C}$ (with its natural compactology). Then $(C^\infty(S), \|\ \|, \tau_\mathcal{K})$ is a Saks space where $\|\ \|$ is the supremum norm and $\tau_\mathcal{K}$ is the topology of uniform convergence on the sets of $\mathcal{K}$. In fact, $(C^\infty(S), \|\ \|, \tau_\mathcal{K})$ is even a Saks C*-algebra (and, in particular, complete).

S is said to be regular if $C^\infty(S)$ separated S;

of countable type if $\mathcal{K}$ possesses a countable

basis $\mathcal{K}_1$ (i.e. every $K \in \mathcal{K}$ is subset of some $K_1 \in \mathcal{K}_1$).

We denote by RCPTOL the full subcategory of regular compactological spaces.

2.2. Proposition: Let $(S,\mathcal{K})$ be a compactological space. Then the following statements are equivalent:

1) $(S,\mathcal{K})$ is regular;

2) there is a completely regular topology on S so that $\mathcal{K}$ is the induced compactology;

3) the restriction operator $C^\infty(S) \longrightarrow C(K)$ is surjective for each $K \in \mathcal{K}$.

Each of these conditions is satisfied if $\mathcal{K}$ is of countable type.

Proof: 1) $\Longrightarrow$ 2): the weak topology on S induced by $C^\infty(S)$ satisfies the required condition.

2) $\Longrightarrow$ 3) follows from Tietze's theorem.

3) $\Longrightarrow$ 1): let s,t be distinct points of S. Then $\{s,t\} \in \mathcal{K}$. Let x be the function

$$s \longmapsto 0, \qquad t \longmapsto 1$$

from $\{s,t\}$ into $\mathbb{C}$. Then an extension $\tilde{x}$ of x in $C^\infty(S)$ separates s and t.

Now let S be of countable type - suppose that (K_n) is an increasing basis (i.e. $K_n \subseteq K_{n+1}$ for each n). We verify condition

Choose $K \varepsilon \mathcal{K}$, $x_1 \varepsilon C(K)$. It is no loss of generality to suppose that $K = K_1$ and $\|x_1\| \leq 1$. Then we can choose inductively a sequence (x_n) where $x_n \varepsilon C(K_n)$, $x_{n+1}|_{K_n} = x_n$ and $\|x_n\| \leq 1$ for each n. Then the x_n define an extension of x_1.

The equivalence of 1) and 2) of 2.2 might suggest the conclusion that there is no difference between completely regular spaces and regular compactologies. This is not the case, the point being that if S and S_1 are completely regular spaces, there are more compactological morphisms from S into S_1 as there are topological ones.

2.3. <u>The spectrum of a Saks algebra</u>: If $(A, \| \|, \tau)$ is a commutative Saks algebra with unit, then $M_\gamma(A)$, the γ-spectrum of A, has a natural compactology: the sets of $\mathcal{K}$ are γ-equicontinuous subsets and, for each $K \varepsilon \mathcal{K}$, τ_K is the restriction of the weak topology $\sigma(A'_\gamma, A)$. Then $M_\gamma(A)$ is even regular. It is obvious that the correspondences

$$S \longmapsto C^\infty(S) \quad \text{and} \quad A \longmapsto M_\gamma(A)$$

are functorial i.e. we have constructed functors C^∞ (from the category of compactological spaces into the category CSC* of commutative, Saks C*-algebras with unit) and M_γ (from the category of commutative Saks algebras with unit into RCPTOL).

2.4. <u>Proposition</u>: CSC* and RCPTOL are quasi-dual under the pair of functors C^∞ and M_γ.

Note that 2.4 means that C^∞ and M_γ are right and left adjoint to each other and that the corresponding unit and co-unit (i.e. the generalised Gelfand-Naimark transform and generalised Dirac transformation) are isomorphisms. The proof is exactly that of II.2.2 and II.2.5 with the suitable changes.

We now establish a duality theory for uniform space. The dual object of a uniform space S will be the CoSaks space $U^\infty(S)$ described in 1.4.II. We repeat the details. We denote by $\mathcal{D}$ the family of all bounded, uniformly continuous pseudometrics on S. $U^\infty(S)$ denotes the vector space of all bounded, uniformly continuous complex-valued functions on S. $(U^\infty(S), \| \|)$ is a Banach space (even a C*-algebra). $\mathcal{H}$ denotes the family of all uniformly bounded, uniformly equicontinuous subsets of $U^\infty(S)$. We remark that a set B is in $\mathcal{H}$ if and only if there is a $d \in \mathcal{D}$ so that B is majorised by d (i.e. $|x(s) - x(t)| \leq d(s,t)$ for $x \in B$, $s,t \in S$). The sufficiency of this condition is clear. On the other hand, if $B \in \mathcal{H}$, then

$$d_B : (s,t) \longmapsto \sup\{|x(s) - x(t)| : x \in B\}$$

is a suitable majorant in $\mathcal{D}$.

If we supply each $B \in \mathcal{H}$ with the topology of pointwise (or compact) convergence, then $(U^\infty(S), \| \|, \mathcal{H})$ is a CoSaks space. In addition, it is a commutative algebra with unit and so we can define the CoSaks spectrum $M_C(U^\infty(S))$ to be the subset of the dual of $U^\infty(S)$ consisting of the unit-preserving multiplicative functionals. We regard $M_C(U^\infty(S))$ as a uniform space with the structure of uniform convergence on the sets of $\mathcal{H}$. We can define, in the obvious way, a mapping δ from S into $M_C(U^\infty(S))$ - the generalised Dirac transform.

2.5. Proposition: $M_C(U^\infty(S))$ is a complete uniform space and δ is a uniform isomorphism from S into $M_C(U^\infty(S))$.

Proof: The statement about the completeness of $M_C(U^\infty(S))$ is clear (for example, $M_C(U^\infty(S))$ is closed in the dual of the Co-Saks space $U^\infty(S)$ and this is a complete locally convex space).

The δ-mapping is injective since $U^\infty(S)$ separated S. Now, by its very definition, the uniform structure of $M_C(U^\infty(S))$ is generated by the pseudometrics

$$\tilde{d}_B : (f,g) \longmapsto \sup\{|f(x) - g(x)| : x \in B\}$$

where B runs through $\mathcal{H}$. Now $\tilde{d}_B$ induces the pseudometric

$$d_B : (s,t) \longmapsto \sup\{|x(s) - x(t)| : x \in B\}$$

on S and the uniform structure of S is generated by $\{d_B : B \in \mathcal{B}\}$. For, as we have already seen, each d_B is in $\mathcal{D}$ and, on the other hand, if $d \in \mathcal{D}$ then

$$B := \{s \longrightarrow d(s,t) : t \in S\}$$

is in $\mathcal{H}$ and $d = d_B$.

We shall now show that S is dense in $M_C(U^\infty(S))$ so that we can identify $M_C(U^\infty(S))$ with the uniform completion of S.

2.6. Lemma: Let $f \in M_C(U^\infty(S))$. Then

a) f is positive;

b) if $x \geq 0$, $f(x) = 0$, then $0 \in \overline{x(S)}$;

c) for each $x_1,\ldots,x_n$ in $U^\infty(S)$ and $\varepsilon > 0$ there is an $s \in S$ so that $\sup_{i=1,..,n} |x_i(s) - f(x_i)| < \varepsilon$.

Proof: a) This is standard (for example, we can regard f as an element of the spectrum of the C*-algebra $U^\infty(S)$).

b) If $|x| \geq \varepsilon$ for some $\varepsilon > 0$, then $|f(x)| \geq f(\varepsilon) = \varepsilon$.

c) Put $x := \Sigma\, |x_i - f(x_i)|$. Then we have: $f(x) = 0$ (for $f(x) = \Sigma\, f(|x_i - f(x_i)|) = \Sigma\, |f(x_i) - f(x_i)| = 0$). Hence, by b), for each $\varepsilon > 0$, there is an $s \in S$ so that $x(s) < \varepsilon$.

2.7. Lemma: Let B be a real-valued set in $\mathcal{H}$, f an element of $M_C(U^\infty(S))$ which vanishes on B. Then $\sup B$ is uniformly continuous and $f(\sup B) = 0$.

Proof: We replace B by the set of functions obtained by adding the suprema of finite subsets. It follows easily from the characterisation of uniform equicontinuity given after 2.4 that the new set is still in $\mathcal{H}$. Also if $x,y \in B$, then $f(\sup\{x,y\}) = 0$ (2.6) and so f vanishes on the extended set. But then $\sup B$ is a pointwise limit and its uniform continuity follows from a classical result. Also f vanishes on sup B because its restriction to B is continuous for the topology of pointwise convergence.

2.8. Lemma: Let $f \in M_C(U^\infty(S))$. Then for each $B \in \mathcal{H}$, $\varepsilon > 0$ there is an $s \in S$ so that $|x(s) - f(x)| < \varepsilon$ for each $x \in B$.

Proof: If $x \in B$, put

$$y_x := \inf (1, |x - f(x)|)$$

Then $B_1 := \{y_x : x \in B\}$ satisfies the conditions of 2.7 and so f vanishes on sup B. This implies the result.

2.9. Proposition: The Dirac transformation is a uniform isomorphism from S onto a dense subspace of $M_C(U^\infty(S))$.

2.10. Corollary: The Dirac transformation is an isomorphism from S onto $M_C(U^\infty(S))$ if and only if S is complete.

Hence, in general, the above method yields a functional analytic construction of the completion of a uniform space. The above results establish a duality between the category of complete uniform spaces and a full subcategory of the category of CoSaks star algebras - namely those which have the form $U^\infty(S)$ for some uniform space S. Unfortunately, in contrast to the situation for compactologies, there is no indication of how one could give an internal characterisation of such algebras.

The analogy between the above results and those of Chapter II make it natural to consider the dual of $(U^{\infty}(S), \| \|, H)$ as the natural space of measures on S (natural in the sense that it is intimately connected with the uniform structure of S). We denote this space by $M^{\infty}(S)$. As the dual of a CoSaks space, it has a natural Saks space structure $(\| \|, \tau_H)$. We denote the corresponding locally convex structure by β_H. Note that $M^{\infty}(S)$ is a subspace of $M(\beta S)$, the space of Radon measures on βS, the Stone-Čech compactification of S (i.e. the Banach algebra spectrum of $U^{\infty}(S)$).

2.11. <u>Proposition</u>: 1) $(M^{\infty}(S), \beta_H)$ is a complete locally convex space;

2) the β_H-bounded subsets of $M^{\infty}(S)$ are precisely the norm-bounded sets;

3) a linear mapping from $M^{\infty}(S)$ into a locally convex space is β_H-continuous if and only if its restriction to the unit ball of $M^{\infty}(S)$ is β_H-continuous;

4) the dual of $(M^{\infty}(S), \beta_H)$ is $U^{\infty}(S)$.

2.12. <u>Remarks</u>: I. Once again, one can relate the uniform properties of S to the linear-topological properties of $M^{\infty}(S)$. As simple examples of this phenomenom, we have:

a) S is discrete if and only if $M^{\infty}(S)$ is a Banach space (and in this case $M^{\infty}(S)$ is just $\ell^1(S)$);

b) S is pseudo-compact (i.e. every uniformly continuous function on S is bounded) if and only if $M^{\infty}(S)$ is semireflexive (or a Schwartz space);

c) S is finite if and only if $M^{\infty}(S)$ is nuclear.

II. As well as the above CoSaks structure on $U^{\infty}(S)$, there are natural Saks space structures which can be defined in a manner similar to that used in Chapter II. For example, we can consider the auxiliary topologies τ_{K} and τ_{B} (uniform convergence on compact, resp. bounded subsets of S resp.). The corresponding dual space will be spaces of measures on S, the corresponding spectra will represent extensions of S.

2.13. Remarks on completely regular spaces: Every completely regular space has a natural uniform structure - the fine structure. In fact, we can identify the category of completely regular spaces with the full subcategory of fine uniform spaces (a uniform space is fine if it possesses the finest uniformity compatible with its associated topology). Hence the above constructions can be applied to completely regular spaces so that the theory of Chapter II could be regarded as a special case of a theory for uniform spaces.

We note that for a completely regular space, the measure space $M^{\infty}(S)$ introduced here coincides with the space $M^{\infty}(S)$ defined in § II.5. In fact, every CoSaks space E has a natural locally convex structure - the finest which agrees with τ_B on each $B \in \mathcal{B}$. It is clear that the locally convex dual of E coincides with its CoSaks dual. In the case of $C^{\infty}(S) = U^{\infty}(S)$ (S a completely regular space, respectively, a fine uniform space), the topology β_∞ defined in II.5 can be shown to be exactly the locally convex structure associated with the CoSaks structure on $C^\infty(S)$.

2.14. Remark: In the light of the above duality for compactological spaces, it is natural to consider the γ-dual of $C^\infty(S)$ (S a regular compactological space). Since the definition of tight measures on a completely regular space involves only its compactology, the definition of II.3.1 can be carried over to define $M_t(S)$ and, once again, $M_t(S)$ is identifiable with the γ-dual of $C^\infty(S)$. However, it can easily be seen that the space of tight measures on the compactological space S is the same as the space of tight measures on the associated completely regular space (i.e. S with the weak topology defined by $C^\infty(S)$). One important point to keep in mind is the fact that if S is a completely regular space which we regard also as a compactological space, then, although the corresponding spaces of tight measures are the same, the weak topologies need not coincide (and indeed this will be the case exactly when S is a $k_{\mathbb{R}}$-space).

A.3. SOME FUNCTORS

3.1. The functors A_t and M_t: In 2.14 we constructed a functor M_t from RCPTOL into MW which takes a compactological space S into the space $M_t(S)$ of tight measures on S ($M_t(S)$ is given a CoSaks structure as the dual of $C^\infty(S)$). There is also a natural functor from MW into RCPTOL, defined as follows: we associate to the CoSaks space $(E,\|\ \|,B)$ the unit ball $B_{\|\ \|}$ of $(E,\|\ \|)$ with the compactology defined by the sets of B which are contained in $B_{\|\ \|}$. This correspondence is functorial and we denote the functor by A_t.

3.2. <u>Proposition</u>: M_t is a left adjoint to A_t.

<u>Proof</u>: Let S be a regular compactological space, E a CoSaks space. We must establish a natural isomorphism between

$$\text{Hom}\ (M_t(S),E) \quad \text{and} \quad \text{Hom}\ (S,A_t(E)).$$

Clearly, any $T \in \text{Hom}\ (M_t(S),E)$ defines an element of $\text{Hom}\ (S,B_{\|\ \|})$ by restriction (we regard S as a subset of $M_t(S)$ by the δ-transformation). On the other hand, if x is a CPTOL-morphism from S into $B_{\|\ \|}$ we can extend x to a morphism $T_x : M_t(S) \longrightarrow E$ by defining

$$T_x(\mu) = \int x\, d\mu$$

where $\int x\, d\mu$ is the weak integral defined by the formula

$$f(\int x\, d\mu) = \int f(x)\, d\mu \qquad (f \in E').$$

3.3. <u>The functors</u> A^∞ <u>and</u> M^∞: Similarly, we can define a pair of functors

$$M^\infty : \text{UNIF} \longrightarrow \text{CSS}$$

$$A^\infty : \text{CSS} \longrightarrow \text{UNIF}$$

where M^∞ ascribes to a uniform space S the measure space $M^\infty(S)$ described in § 2 and A^∞ associates to a Saks space its unit ball with the uniform structure induced by τ.
As for 3.2, one can prove:

3.4. Proposition: M^∞ is a left adjoint for A^∞.

3.5. Remarks: I. We can, as usual, deduce certain continuity properties of the above functors from the adjointness.
II. If we compare 3.4 with the corresponding result for Banach spaces (see BUCHWALTER [8], § I.2.20) we see that M^∞ is a continuous analogue of the ℓ^1-functor. In particular, every complete Saks space is a quotient of an $M^\infty(S)$-space.
III. Proposition 3.2 and 3.4 can be interpreted as stating that the injections $S \longrightarrow M^\infty(S)$ and $S \longrightarrow M_t(S)$ have certain universal properties (which we shall not make explicit here), in other words, that $M^\infty(S)$ and $M_t(S)$ are the free Saks (resp. CoSaks) spaces over S.

A.4. DUALITY FOR SEMIGROUPS AND GROUPS

In § A.2 we have extended the Gelfand-Naimark duality for compact spaces to regular compactologies. HOFMANN has given a duality theory for compact semigroups and groups based on Gelfand-Naimark duality and we are now in a position to extend this duality to a large class of (not necessary abelian) groups and semigroups which includes locally compact groups and semigroups.

4.1. Definition: A compactological semigroup is a semigroup in the category of compactological spaces i.e. it is a pair (S,m) where S is a compactological space and m is a multiplication on S under which S is a semigroup so that m is a compactological morphism from $S \times S$ into S. The compactological semigroups form a category (the morphisms are the compactological morphisms which preserve multiplication).
A compactological group is a compactological semigroup (S,m) with unit so that (S,m) is a group and inversion is a compactological morphism. A compactological group (semigroup) is regular if its underlying compactology is regular. We can thus form three categories

SGRCPTOL $\quad$ $\mathrm{SGRCPTOL}_e$ $\quad$ and $\quad$ GRCPTOL

whose objects are, respectively, regular compactological semigroups, semigroups with unit and groups.

In order to motivate the following definition, we reformulate that of a compactological group in terms of commutative diagrams: a compactological group is a quadruple (S,m,e,i) where S is a compactological space and

$$m : S \times S \longrightarrow S, \quad e : \{\cdot\} \longrightarrow S \quad \text{and} \quad i : S \longrightarrow S$$

are CPTOL-morphisms so that the following diagrams commute:

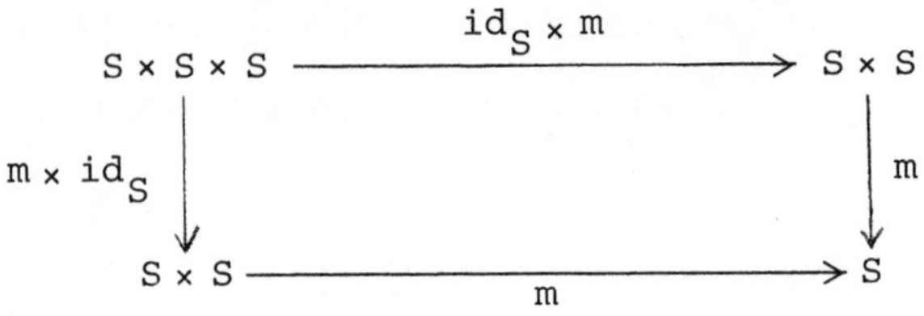

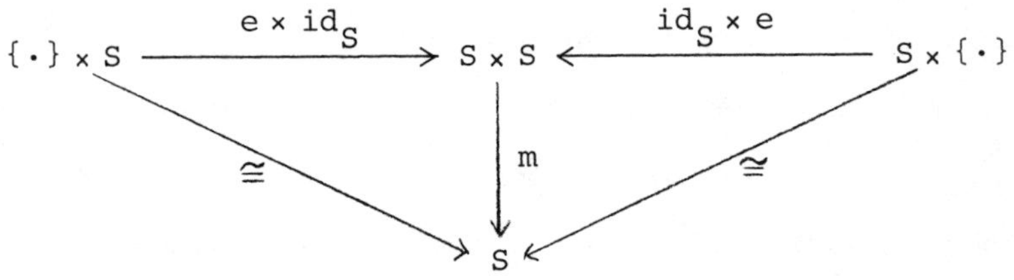

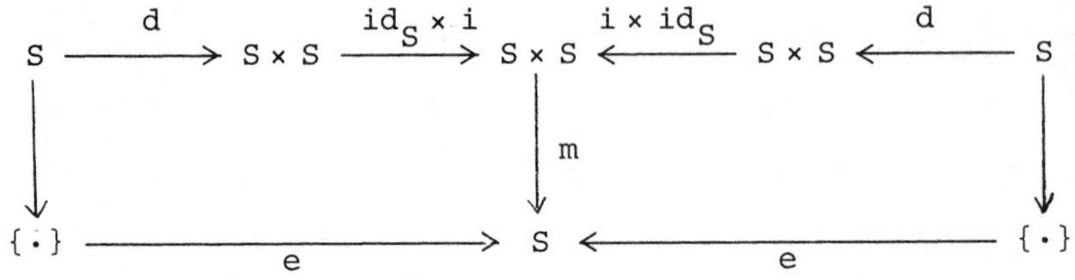

({·} is the one-point set i.e. the final object in CPTOL, d is the diagonal mapping).

In order to describe the dual objects for the above categories, we simply reverse the arrows in CSC* (the category of commutative Saks C*-algebras with unit).

4.2. <u>Definition</u>: A <u>Saks C*-cogebra</u> is a commutative, complete Saks C*-algebra with unit $(E, \| \ \|, \tau)$ together with a CSC*-morphism $c : E \longrightarrow E \hat{\hat{\otimes}}_{\gamma} E$ so that

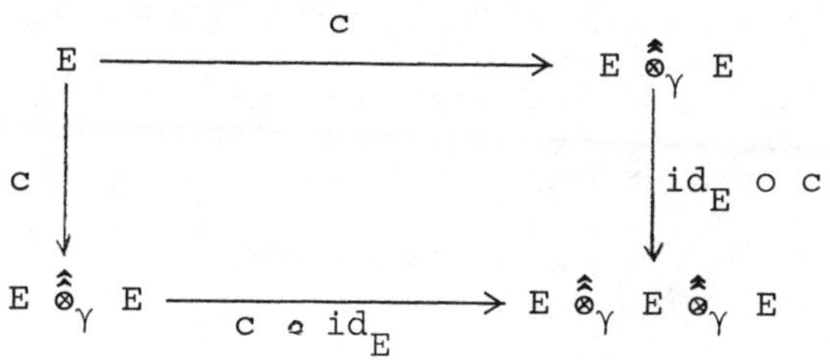

commutes.

A <u>co-unit</u> for a Saks C^*-cogebra is a morphism $\eta : E \longrightarrow \mathbb{C}$ so that

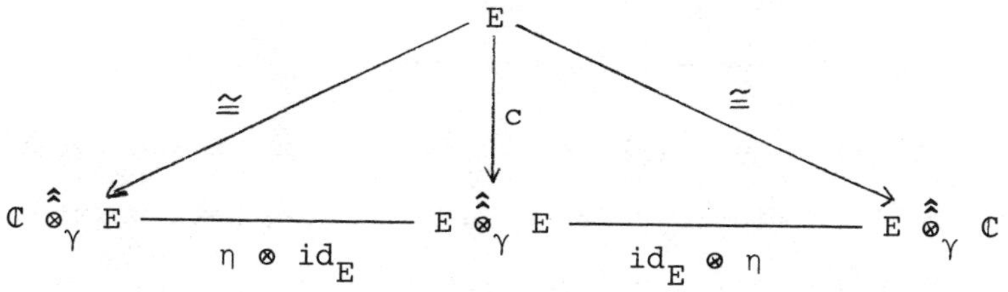

commutes.

A Saks <u>C^*-cogroup</u> is a Saks C^*-cogebra with co-unit together with a morphism $a : E \longrightarrow E$ (the antipodal mapping) so that the diagram

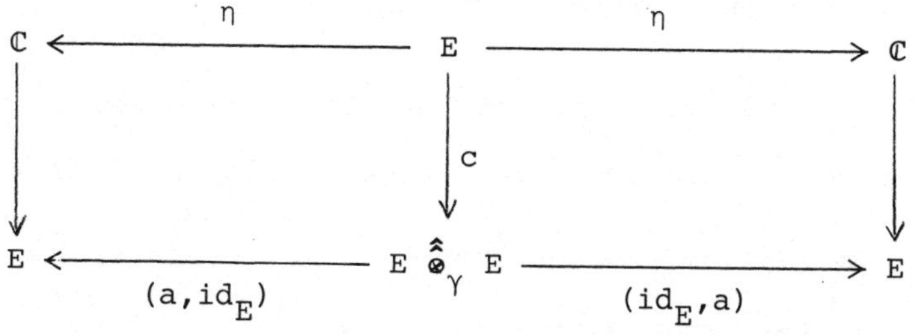

((a,id_E) denotes the canonical morphism from $E \hat{\hat{\otimes}}_\gamma E$, the co-product in CSC^* into E, defined by the mappings a and id_E).

We denote by SC^*COG, SC^*COG_C and SC^*COGR the categories of Saks C^*-cogebras, Saks C^*-cogroups with co-unit and Saks C^*-semigroups resp. We leave to the care of the reader the task of untangling the suitable morphisms.

4.3. Examples: If (S,m,e,i) is a compactological group, then $C^\infty(S)$ is a Saks C^*-cogroup under the operations

$$c : x \longmapsto ((s,t) \longmapsto x(st))$$
$$\eta : x \longmapsto x(e)$$
$$a : x \longmapsto (s \longmapsto x(s^{-1})).$$

i.e. $c = C^\infty(m)$, $\eta = C^\infty(e)$, $a = C^\infty(a)$. (Here we are reverting to the normal terminology for groups i.e. we have written st for $m(s,t)$, e for $e(\cdot)$, s^{-1} for $i(s)$).

Similar remarks hold for semigroups (with a unit). It is clear that the functor C^∞ lifts to a functor from

$$\begin{array}{lll} \text{SGRCPTOL} & \text{into} & SC^*COG \\ \text{SGRCPTOL}_e & \text{into} & SC^*COG_C \\ \text{GRCPTOL} & \text{into} & SC^*COGR. \end{array}$$

On the other hand, if (E,c,η,a) is a Saks C^*-cogroup, then $M_\gamma(E)$ is a compactological group under the operations

$$m : (f,g) \longmapsto (x \longmapsto f \otimes g(c(x)))$$
$$e : \cdot \longmapsto \eta$$
$$a : f \longmapsto (x \longmapsto f(a(x))).$$

Then we can regard M_γ as a functor from

SC*COG into SGRCPTOL

SC^*COG_C into $SGRCPTOL_e$

SC*COGR into GRCPTOL.

Combining 2.4 with the above remarks, we get the following duality theorem:

4.4. Proposition: The functors C^∞ and M_γ induce a duality between

SC*COG and SGRCPTOL

SC^*COG_C and $SGRCPTOL_e$

SC*COGR and GRCPTOL.

4.5. Remark: The locally compact groups (semigroups etc.) can be regarded as a full subcategory of GRCPTOL etc. so that 4.4 contains a duality theory for locally compact groups, semigroups and semigroups with unit. The dual objects are recognised by their perfectness (cf. II.2.6). Of course, every topological group has a natural compactology under which it is a regular compactological group. However, one cannot get a duality for topological groups from 4.4 (the problem lies once again in the fact that there are more compactological than topological group morphisms between two topological groups). One can get a duality theory if one is prepared to accept as morphisms those group morphisms which are continuous on compacta but this is of course just another way of saying 4.4 and it seems better to use the language of compactologies from the beginning.

4.6. <u>The Bohr compactification</u>: As an example of the usefulness of this duality we give a (conceptually, at least) very simple construction of the Bohr compactification. The reader will observe that it is formally exactly the same as the construction of the Stone-Čech compactification given in § II.2 where, we recall, we formed the dual object $(C^{\infty}(S), \| \|, \tau_K)$ of S, forgot the mixed structure and passed over to the dual object $M(C^{\infty}(S))$ of the Banach algebra $C^{\infty}(S)$. In the case of groups we require an intermediary step since it is not quite so easy to forget the mixed structure of a Saks cogroup.

Essentially the same construction works for semigroups (with or without unit) or groups. For the sake of simplicity of notation, we consider semigroups.

Let (E,c) be a Saks C^*-cogebra. In general, the compact space M(E) (i.e. $\beta(M_\gamma(E))$) is not a semigroup. The problem lies in the fact that the associated C^*-algebra E need not be a cogebra (since c(E) need not be contained in the C^*-tensor product $E \hat{\hat{\otimes}} E$ which is, in general, smaller than $E \hat{\hat{\otimes}}_\gamma E$). We can circumvent this difficulty with the following Lemma:

4.7. <u>Lemma</u>: Let (E,c) be a Saks C^*-cogebra. Then there is a largest C^*-subalgebra $\tilde{E}$ of E with the property that $c(\tilde{E}) \subseteq \tilde{E} \hat{\hat{\otimes}} \tilde{E}$. $\tilde{E}$, with the induced norm, is a C^*-cogebra. The assignment $E \longrightarrow \tilde{E}$ is functorial.

Proof: For each ordinal α we define a subalgebra E_α of E inductively by

$$E_o := E$$
$$E_\alpha := c^{-1}(E_\beta \hat{\hat{\otimes}} E_\beta) \qquad (\alpha = \beta + 1)$$
$$E_\alpha := \bigcap_{\beta<\alpha} E_\beta \qquad (\alpha \text{ a limit ordinal}).$$

Then the family $\{E_\alpha\}$ is eventually stationary. We denote its limit by $\widetilde{E}$. Then $\widetilde{E}$ is a C*-subalgebra of E with the desired properties.

The functor $U : E \longmapsto \widetilde{E}$ is now the required forgetful functor from SC*COG into C*COG, the category of C*-cogebras. We can now define a functor

$$B := M \circ U \circ C^\infty$$

from SGRCPTOL into SGCOMP, the category of compact semigroups. If S is a compactological semigroup, B(S) is called the Bohr compactification of S. There is a natural morphism $j_S : S \longrightarrow B(S)$, the adjoint of the embedding from $\widetilde{C^\infty(S)}$ into $C^\infty(S)$. j_S has dense image but need not be injective.

4.8. Proposition: B(S) has the following universal property: every morphism from S into a compact semigroup T factorises over j_S.

Proof: If $\phi : S \longrightarrow T$ is a morphism, then

$$C^\infty(\phi) : C(T) \longrightarrow C^\infty(S)$$

is a SC*COG-morphism. Since $UC(T) = C(T)$ and U is a functor, acting on morphisms by restricting them, we see that $C^{\infty}(\phi)$ maps $C(T)$ into $\widetilde{C^{\infty}(S)}$ and the result follows by dualising.

4.9. Remarks: I. Almost the same construction produces Bohr compactification functors for compactological semigroups with unit or compactological groups. We remark that in the case of groups, in the proof of Lemma 4.7 we replace c by the mapping $\tilde{c} := (\mathrm{id}_E \otimes a) \circ c$ to obtain the C^*-cogroup $\tilde{E} \subseteq E$.

II. As far as we know, this construction of the Bohr compactification is more general than those which have appeared until now in the literature. If one compares the above construction with the standard ones, then one can see that $\widetilde{C^{\infty}(S)}$ can usually be concretely identified with a space of almost periodic functions on S.

We close this section with some remarks on the convolution algebra of a semigroup and on PONTRYAGIN duality in the light of the above duality theory.

If (E,c) is a Saks C^*-cogebra then the dual E'_{γ} of E has a natural multiplication, defined as follows: if $f,g \in E'_{\gamma}$ then $f * g$ is the form

$$x \longmapsto f \otimes g\ (c(x)).$$

Then E'_{γ} is a Banach algebra under this multiplication. It is commutative if c is co-commutative i.e. if the diagram

$$\begin{array}{ccc} E \,\hat{\hat{\otimes}}_\gamma\, E & \longrightarrow & E \,\hat{\hat{\otimes}}_\gamma\, E \\ & {}_{c}\nwarrow \quad \nearrow^{c} & \\ & E & \end{array}$$

commutes where the horizontal arrow denotes the flip operator $(x \otimes y \longmapsto y \otimes x)$. If E is the dual object $C^\infty(S)$ of the semigroup S then E is co-commutative if and only if S is commutative.

Now if E is the cogebra $C^\infty(S)$ (S a compactological semigroup) then E'_γ is $M_t(S)$, the space of bounded Radon measures on S. The Banach algebra structure of $M_t(S)$ described above is that of ordinary convolution. For if $x \in C^\infty(S)$, $\mu,\nu \in M_t(S)$, then

$$\begin{aligned} \int x \, d\,(\mu * \nu) &= (\mu \otimes \nu)(c(x)) \\ &= \int x(st)\, d\,\mu \otimes \nu\,(s,t) \\ &= \int x(st)\, d\,\mu(s)\, d\,\nu(t). \end{aligned}$$

Now suppose that (E,c,η,a) is an object of SC*COGR. $x \in E$ is called <u>strongly primitive</u> if

a) $c(x) = x \otimes x$;

b) $\eta(x) = 1$;

c) $a(x) = x^{-1}$ (the inverse of x in E).

We denote by P(E) the set of strongly primitive elements of E.

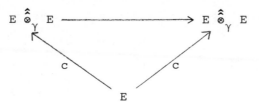

4.10. <u>Proposition</u>: $P(E)$, with the topology and multiplicative structure induced from (E,γ), is a topological group. It is a subset of the unit sphere of E.

<u>Proof</u>: $P(E)$ is closed under multiplication for if $x,y \in P(E)$, then

a) $c(xy) = c(x)c(y) = (x \otimes x)(y \otimes y) = xy \otimes xy$.

b) $\eta(xy) = \eta(x)\eta(y) = 1.1 = 1$.

c) $a(xy) = a(x)a(y) = x^{-1}y^{-1} = (xy)^{-1}$ (E is commutative).

The constant function 1 is a unit for $P(E)$.
If $x \in P(E)$, then $a(x) = x^{-1}$ is an inverse of x. Hence $P(E)$ is a topological group.

Finally, since $\|x\| \|x\| = \|x \otimes x\| = \|c(x)\| \leq \|x\|$ and $1 = \|1\| = \|\eta(x)\| \leq \|x\|$ we conclude that $\|x\| = 1$ for all $x \in P(E)$.

4.11. <u>Definition</u>: Let S be a commutative compactological group. A character on S is a GCPTOL-morphism from S into the circle group T. The set $\hat{S}$ of all characters form a group. $\hat{S}$, with the topology of uniform convergence on the compacta of S, is a topological group (in fact, a complete topological group). If S is locally compact, then this notion coincides with the classical one.

4.12. <u>Proposition</u>: Let S be a commutative, compactological group. Then $\hat{S} = P(C^{\infty}(S))$ (as topological groups).

Proof: If $x \in P(C^{\infty}(S))$, then $\|x\| = 1 = \|x^{-1}\|$ and so x takes its values in T. It is trivial to verify that the conditions a) - c) of the definition before 4.10 ensure that x is a group morphism. Since $\gamma = \tau_K$ on $B_{\| \|}$, it follows that the topologies coincide.

4.13. Remark: Using the duality of 2.10 we could presumably establish a duality theory for semigroups and groups in the category of uniform spaces (note that a semigroup in UNIF is just a topological semigroup while a group in UNIF is a topological group in which the left and right uniformities coincide). For this purpose it would be necessary to identify a tensor product in MW which defines the coproduct on the category dual of UNIF.

A.5. EXTENSIONS OF CATEGORIES

A central theme of this book could be regarded as the attempt to extend a classical theory for some restricted category to a larger category (for example, the Gelfand-Naimark duality between compact spaces and commutative C^*-algebras with unit was extended to a duality between compactologies and Saks algebras). In each case, the extended category could be regarded as a kind of completion of the original category in the sense that its objects were projective (or inductive) limits of spectra in the smaller category. Of course there is a general construction lurking in the background and we describe this now.

5.1. Spectra over categories: Let $\mathcal{A}$ be a category, A a directed set regarded as a category in the usual way.

A spectrum in $\mathcal{A}$ over A is a covariant functor i from A into $\mathcal{A}$. We write $(i : A \longrightarrow \mathcal{A})$ for such a spectrum. If $\alpha \le \beta$ it will sometimes be convenient to write E_α for $i(\alpha)$ and $i_{\alpha\beta}$ for the morphism from E_α into E_β. We shall thus sometimes write the spectrum more explicitly in the form

$$\{i_{\alpha\beta} : E_\alpha \longrightarrow E_\beta,\ \alpha,\beta \in A,\ \alpha \le \beta\}$$

If $E = \{i_{\alpha\beta} : E_\alpha \longrightarrow E_\beta,\ \alpha,\beta \in A\}$ and $F = \{j_{\gamma\delta} : F_\gamma \longrightarrow F_\delta,\ \gamma;\delta \in B\}$ are spectra over $\mathcal{A}$, a pre-mapping from E into F is a pair $(r,\{T_\alpha\})$ where r is an increasing map from A into B and T_α is a morphism from E_α into $F_{r(\alpha)}$ so that, for $\alpha \le \beta$, the diagram

$$\begin{array}{ccc} E_\alpha & \xrightarrow{\quad i_{\alpha\beta}\quad} & E_\beta \\ {\scriptstyle T_\alpha}\big\downarrow & & \big\downarrow{\scriptstyle T_\beta} \\ F_{r(\alpha)} & \xrightarrow[\quad J_{r(\alpha),r(\beta)}\quad]{} & F_{r(\beta)} \end{array}$$

commutes.

If $E\ (= (i : A \longrightarrow \mathcal{A}))$, $F\ (= (j : B \longrightarrow \mathcal{A}))$ and $G\ (= (k : \Gamma \longrightarrow \mathcal{A}))$ are spectra in $\mathcal{A}$ and $(r,\{T_\alpha\})$ (resp. $(s,\{U_\beta\})$) is a premapping from E into F (resp. from F into G) then we define the composition of $(r,\{T_\alpha\})$ and $(s,\{U_\beta\})$ to be the premapping $(t,\{V_\alpha\})$ where

$$t := s \circ r \qquad \text{and} \qquad V_\alpha := U_{r(\alpha)} T_\alpha .$$

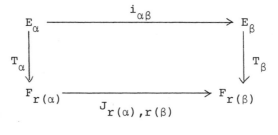

We have clearly defined a category whose objects are spectra in $\mathcal{A}$ and whose morphisms are premappings. The category that we require is a quotient of this category.

Two premappings $(r,\{T_\alpha\})$ and $(r',\{T'_\alpha\})$ are equivalent if

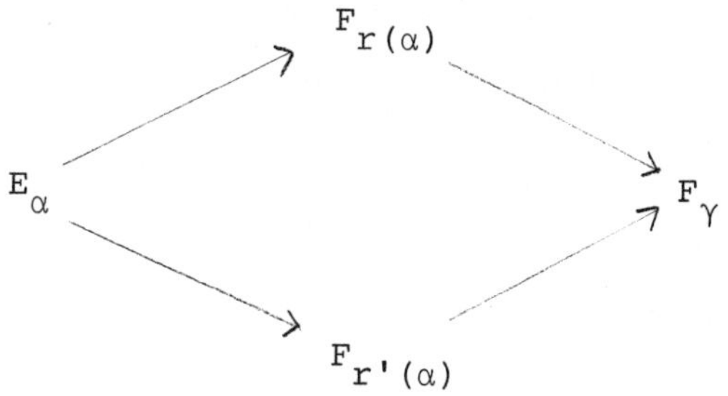

commutes for each α and each γ with $\gamma \geq r(\alpha)$ and $\gamma \geq r'(\alpha)$. It can then easily be seen that this is, in fact, an equivalence relation. A <u>mapping between</u> two spectra is an equivalence class of premappings. One can check that this equivalence relation is compatible with composition so that compositions of mappings are well-defined. Hence we have a new category whose objects are spectra in $\mathcal{A}$ and whose morphisms are mappings. We denote it by Spec($\mathcal{A}$).

Dually, we can define cospectra over $\mathcal{A}$ as contravariant functors from A into $\mathcal{A}$. These are denoted by $(\pi : A \longrightarrow \mathcal{A})$ or, more explicitly, by

$$\{\pi_{\beta\alpha} : E_\beta \longrightarrow E_\alpha,\ \alpha,\beta \in A,\ \alpha \leq \beta\}$$

The corresponding category is denoted by Cospec($\mathcal{A}$).

If, in the above construction, we consider only spectra over a fixed directed set A we obtain categories $\mathrm{Spec}_A(\mathcal{A})$ and $\mathrm{Cospec}_A(\mathcal{A})$. In the special case where $A = \mathbb{N}$ with its usual ordering, we write $\mathrm{Spec}_\omega(\mathcal{A})$, $\mathrm{Cospec}_\omega(\mathcal{A})$.

We can regard $\mathcal{A}$ as a full subcategory of $\mathrm{Spec}(\mathcal{A})$ by identifying $A \in \mathcal{A}$ with the natural spectrum over a one-point set.

5.2. Proposition: 1) $\mathrm{Spec}(\mathcal{A})$ has inductive limits;
2) $\mathcal{A}$ is dense in $\mathrm{Spec}(\mathcal{A})$ (i.e. every object in $\mathrm{Spec}(\mathcal{A})$ is the inductive limit of a spectrum in $\mathcal{A}$);
3) if $\mathcal{A}$ has finite sums and co-equalisers, then so does $\mathrm{Spec}(\mathcal{A})$ Hence in this case $\mathrm{Spec}(\mathcal{A})$ is complete.

The proof of 5.2 is simple but rather tedious to write out specifically. Of course, there is a dual version for $\mathrm{Cospec}(\mathcal{A})$.

5.3. Examples: I. Let COMP denote the category of compact (Hausdorff) spaces. We show that Spec(COMP) is naturally identifiable with CPTOL, the category of compactological spaces. To do this, we construct functors

$$F : \mathrm{Spec}(\mathrm{COMP}) \longrightarrow \mathrm{CPTOL}, \quad G : \mathrm{CPTOL} \longrightarrow \mathrm{Spec}(\mathrm{COMP})$$

as follows: if $\mathbb{K} = \{i_{\alpha\beta} : K_\alpha \longrightarrow K_\beta,\ \alpha \le \beta,\ \alpha,\beta \in A\}$ is a spectrum of compacta, we denote by $F(\mathbb{K})$ its inductive limit in CPTOL. On the other hand, if $(S,\mathcal{K})$ is a compactological space we define $G(S)$ to be the spectrum

$$\{i_{K,K_1} : K \longrightarrow K_1,\ K,K_1 \in \mathcal{K},\ K \subseteq K_1\}$$

in Spec(COMP). It is clear that F and G are functorial and that they define an isomorphism from Spec(COMP) onto CPTOL (i.e. $F \circ G$ and $G \circ F$ are naturally equivalent to the appropriate identity functors). We remark here that Spec(COMP) is strictly larger than COMP despite the fact that inductive limits exist in COMP. This is generally true and is due to the fact that morphisms between inductive limits of spectra say in COMP do not factorise over the components as in CPTOL.

In the following examples we shall identify Spec(A) and Cospec(A) for concrete A without specifying the details as above. In each case, it is clear how the isomorphisms are constructed.

II. The category CLOCCONV of complete locally convex spaces can be identified with Cospec(BAN). This result follows from the well known fact that a complete locally convex space has a canonical representation as a projective limit of a spectrum of Banach spaces so that every morphism into a Banach space factorises over an element of the spectrum.

III. We have the following identities between categories of Saks spaces:

CSS = Cospec(BAN_1)

CSA = Cospec(BALG)

CSC*= Cospec(CC*)

where CSA, BALG, CSC* and CC* denote the categories of complete

Saks algebra, Banach algebras, commutative Saks C^*-algebras and commutative C^*-algebras respectively.

IV. The category of complete convex bornological spaces can be identified with Spec(BAN). The category MW can be regarded as Spec(W) where W is the category of Waelbroeck spaces (i.e. the subcategory of CSS consisting of those objects $(E,\| \|,\tau)$ with $B_{\| \|}$ τ-compact - see BUCHWALTER [8]).

V. The category of complete uniform space is identifiable with Cospec(M) where M is the category of metric spaces (with contractions as morphisms).

5.3. <u>Extensions of functors</u>: Let $\mathcal{A}$ and $\mathcal{B}$ be categories, F a covariant functor from $\mathcal{A}$ into $\mathcal{B}$. If $i : A \longrightarrow \mathcal{A}$ is a spectrum in $\mathcal{A}$ over A, then $F \circ i$ is a spectrum in $\mathcal{B}$ over A. If $(r,\{T_\alpha\})$ is a premapping from a spectrum $(i : A \longrightarrow \mathcal{A})$ into a spectrum $(i_1 : A_1 \longrightarrow \mathcal{A})$ then $(r,\{F(T_\alpha)\})$ is a premapping from $F \circ i$ into $F \circ i_1$. The association

$$(r,\{T_\alpha\}) \longmapsto (r,\{F(T_\alpha)\})$$

preserves equivalence of premappings and composition and so F induces a functor, which we denote by Spec(F), from Spec($\mathcal{A}$) into Spec($\mathcal{B}$). Similary, F induces a (covariant) functor Cospec(F) from Cospec($\mathcal{A}$) into Cospec($\mathcal{B}$).

Of course a similar construction can be carried out for a contravariant functor G from $\mathcal{A}$ into $\mathcal{B}$. Then G induces functors

Spec(G) from Spec($\mathcal{A}$) into Cospec($\mathcal{B}$)

and Cospec(G) from Cospec($\mathcal{A}$) into Spec($\mathcal{B}$).

5.4. Examples: Amongst many possible examples, we mention only that the duality functors from CPTOL into CSC* (resp. from CSS into MW) discussed earlier in this Chapter are the extensions of the functors from COMP into CC* (resp. from BAN_1 into W) in the sense of 5.3.

5.5. Lemma: Let E be an object of $\mathcal{A}$, $\mathbb{F} = \{i_{\alpha\beta} : F_\alpha \longrightarrow F_\beta\}$ a spectrum in $\mathcal{A}$ over A. Then $\{Hom_{\mathcal{A}}(E,F_\alpha)\}_{\alpha \varepsilon A}$ forms an inductive spectrum in SET and $Hom_{spec(\mathcal{A})}(E,\mathbb{F})$ is naturally identifiable with its inductive limit.

Proof: We remark first that composition by $i_{\alpha\beta}$ induces a mapping from $Hom(E,F_\alpha)$ into $Hom(E,F_\beta)$ and it is under these linking mappings that $\{Hom_{\mathcal{A}}(E,F_\alpha)\}$ forms an inductive spectrum. If $\alpha \varepsilon A$ and $T_\alpha \varepsilon Hom_{\mathcal{A}}(E,F_\alpha)$ then T_α defines a premapping from E into $\mathbb{F}$ and every premapping has this form (recall that we are identifying E with the spectrum $\cdot \longrightarrow E$ over the one-point set). Hence there is a natural mapping from the disjoint union $\coprod_{\alpha \varepsilon A} Hom_{\mathcal{A}}(E,F_\alpha)$ onto the set of premappings from E into $\mathbb{F}$. Now the SET-inductive limit is the quotient of this direct union under the equivalence relation:

$$T_\alpha \sim U_\beta \text{ if and only if there is a } \gamma \geq \beta,\alpha \text{ so that } i_{\alpha\gamma} \circ T_\alpha = i_{\beta\gamma} \circ U_\beta$$

and for premappings of this special form this is just the equivalence relation which defines morphisms in Spec(A).

5.6. Lemma: Let $\mathbb{E} = \{i_{\alpha\beta} : E_\alpha \longrightarrow E_\beta\}$, $\mathbb{F}$ be objects of Spec(A). Then $\{\mathrm{Hom}_{\mathrm{Spec}(A)}(E_\alpha, \mathbb{F})\}$ forms a projective spectrum in SET and $\mathrm{Hom}_{\mathrm{Spec}(A)}(\mathbb{E},\mathbb{F})$ is naturally identifiable with its projective limit.

Proof: We note that $\{i_{\alpha\beta} : E_\alpha \longrightarrow E_\beta\}$ can be regarded as an inductive spectrum in Spec(A) and $\mathbb{E}$ is its inductive limit there. Hence the result is merely a restatement of the universal property of inductive limits.

5.7. Proposition: Let $\mathbb{E}$ and $\mathbb{F}$ be objects of Spec(A). Then

$$\mathrm{Hom}_{\mathrm{Spec}(A)}(\mathbb{E},\mathbb{F}) = \varprojlim_{\alpha} \varinjlim_{\gamma} \mathrm{Hom}_A(E_\alpha, F_\gamma)$$

5.8. Proposition: Let $G : A \longrightarrow B$ and $H : B \longrightarrow A$ be an adjoint pair (with G adjoint on the left). Then Spec(G) is left-adjoint to Spec(H).

Proof: We have an identification

$$\mathrm{Hom}_B(G(E),F) = \mathrm{Hom}_A(E,H(F))$$

for each object E of A and object F of B, the identification being natural in E and F. Now we have, for

$$\mathbb{E} = \{i_{\alpha\beta} : E_\alpha \longrightarrow E_\beta,\ \alpha,\beta \in A\}$$
$$\mathbb{F} = \{j_{\gamma\delta} : F_\gamma \longrightarrow F_\delta,\ \gamma,\delta \in B\}$$

the equations

$$\mathrm{Hom}_{\mathrm{Spec}(\mathcal{B})}(\mathrm{Spec}(G)\ \mathbb{E},\mathbb{F}) = \underset{\overleftarrow{\alpha}}{\lim}\ \underset{\overrightarrow{\gamma}}{\lim}\ \mathrm{Hom}_{\mathcal{B}}(G(E_\alpha),F_\gamma)$$

$$= \underset{\overleftarrow{\alpha}}{\lim}\ \underset{\overrightarrow{\gamma}}{\lim}\ \mathrm{Hom}_{\mathcal{A}}(E_\alpha,H(F_\gamma)) = \mathrm{Hom}_{\mathrm{Spec}(\mathcal{A})}(\mathbb{E},\mathrm{Spec}(H)\,\mathbb{F}).$$

Of course, there are numerous variations of 5.7 and 5.8 obtained by permuting Spec and Cospec and introducing contravariant adjointness. They can all be obtained from the above results by noting e.g. that $\mathrm{Cospec}(\mathcal{A}) = (\mathrm{Spec}(\mathcal{A}^{op}))^{op}$.

5.9. Remark: Using similar techniques, it would presumably be possible to give an abstract duality result which would include for example the transition from the classical Gelfand-Naimark duality to that given in 2.4 i.e. to show that functors which establish a duality between $\mathcal{A}$ and $\mathcal{B}$ extend to a duality between Spec($\mathcal{A}$) and Cospec($\mathcal{B}$). However, care must be taken since the example CPTOL = Spec(COMP) shows that a separating duality for $\mathcal{A}$ need not induce a separating duality on Spec($\mathcal{A}$). It is presumable necessary to restrict the duality to some reflective subcategory of Spec($\mathcal{A}$) for which the duality is separating (in our example, to the subcategory RCPTOL of CPTOL).

5.10. Remark: Using the construction of Cospec and Spec, together with those of semigroup in and dual to the category, one can construct from a given category a whole range of new categories. For example, the "family trees" of the categories BAN and BAN_1 look like this:

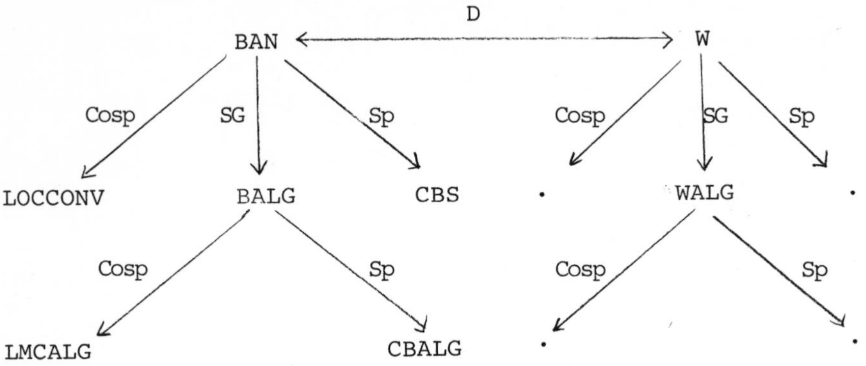

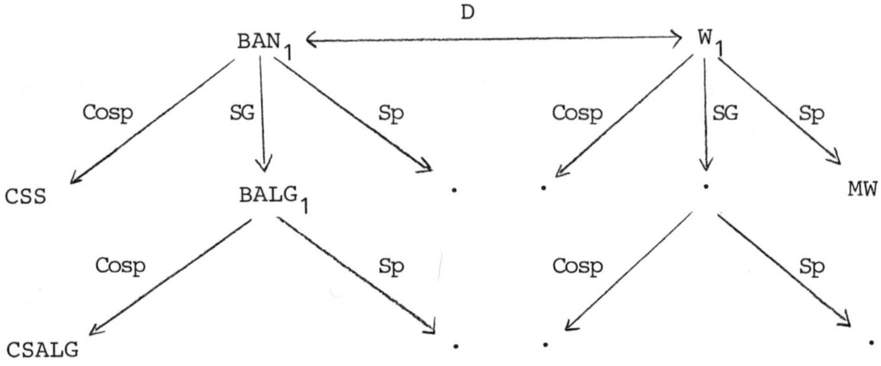

The arrows are labelled by words or initials which indicate the process applied to the source to obtain the target. The labels have the following meaning:

$A \xleftrightarrow{D} B$: A and B are dual categories;

$A \xrightarrow{Sp} B$: $B = \text{Spec}(A)$;

$A \xrightarrow{Cosp} B$: $B = \text{Cospec}(A)$;

$A \xrightarrow{SG} B$: B is the category of semigroups over A.

A subscript "1" means that morphisms are assumed to be contractions.

We have given titles only to those categories which have been mentioned in this monograph. However, many of the nameless categories are of importance.

It is thus remarkable how many of the categories which have proved so important in the recent development of functional analysis can be obtained from (essentially) one category, that of Banach spaces. Similar tables can be constructed with, for example, C^*-algebras, metric spaces or compact spaces as the "mother category".

A.6. NOTES

BUCHWALTER [8], using ideas of WAELBROECK, identified the category dual to that of Banach spaces. We have extended this result in the natural way to Saks spaces. In [9] he also established a duality theory for compactologies which is essentially equivalent to that given in § 2. The latter has the advantage that we can describe explicitly the sum (i.e. the tensor product) in the dual category so that multiplication is preserved by the duality. 2.2 is Théorème 1.1.3 of [9]. For the duality theory for uniform spaces see, for example, PUPIER [34]. Results analogous to 2.11 for completely regular spaces can be found in

ROME [35]. Measures on uniform spaces have been studied in detail by AZZAM [1], BEREZANSKY [6] and DEAIBES [15]. In § 3 we have extended some parts of section III.1 in BUCHWALTER [8] to Saks spaces. § 4 is based on MICHOR [30] and COOPER and MICHOR [14]. The idea of using Gelfand-Naimark duality as a basis for a duality theory for compact groups was used by HOFMANN [20]. In § 5 we have followed COOPER [13]. Similar constructions have appeared elsewhere in the literature (see, for example, MARDEŠIĆ [28]).

As general references for this chapter we recommend Mac LANE [27] SCHUBERT [36] and SEMADENI [37] (category theory) and ISBELL [22] and [23] (uniform spaces). BARR [2] - [4] also considers categories of mixed spaces. The decisive influence of the work of BUCHWALTER on this appendix will be obvious to anyone who is familiar with the contents of [8] and [9].

We take this opportunity to mention two topics which we have not discussed in this monograph. Firstly, there exists a non-commutative version of the theory of chapter II based on the concept of the double centraliser of a Banach algebra (or, more generally, of a Banach module). We have omitted this topic, firstly because of the large number of preliminary definitions that it would require and secondly because we have nothing new to say about it. We refer the reader to [10],[11],[12],[16],[17],[18], [24],[25],[26],[29],[33],[38],[39],[40],[41],[43].

One of the motivations for the introduction of strict topologies has its origins in harmonic analysis (see BEURLING [37] and HERZ [19]) and this is potentially one of the most important fields of application of mixed topologies. The present author does not regard himself as competent to cover this topic. BENEDETTO [5] contains some remarks on applications of the strict topology to spectral synthesis.

REFERENCES FOR THE APPENDIX

[1] N. AZZAM — Mesures sur les espaces uniformes (Prepublication No. 2 - Université de St. Etienne, 1974).

[2] M. BARR — Duality of Banach spaces, Cah. de Top. Geom. Diff. 17 (1976) 15-32.

[3] — Closed categories and topological vector spaces, Cah. de Top. Geom. Diff. 17 (197 223-234.

[4] — Closed categories and Banach spaces (Preprint).

[5] J. BENEDETTO — Spectral synthesis (Stuttgart 1975).

[6] Ia. BEREZANSKY — Measures on uniform spaces and molecular measures, Transl. Moscow Math. Soc. 19 (1968) 1-40.

[7] A. BEURLING — Une théorème sur les fonctions bornées e uniformement continues sur l'axe reel, Acta Math. 77 (1945) 127-136.

[8] H. BUCHWALTER — Topologies, bornologies et compactologie Thèse Doctorat, Lyon 1968.

[9] — Topologies et compactologies, Publ. Dep. Math. Lyon 6-2 (1969) 1-74.

[10] R.C. BUSBY — Double centralizers and extensions of C*-algebras, Trans. Amer. Math. Soc. 132 (1968) 79-99.

[11] H.S. COLLINS and R.A. FONTENOT — Approximate identities and the strict topology, Pac. J. Math. 43 (1972) 63-79.

[12] H.S. COLLINS and W.H. SUMMERS Some applications of Hewitt's factorisation theorem, Proc. Amer. Math. Soc. 21 (1969) 727-733.

[13] J.B. COOPER Remarks on applications of category theory to functional analysis (Preprint, 1974).

[14] J.B. COOPER and P. MICHOR Duality of compactological and locally compact groups (Categorical topology - Springer Lecture Notes 540, 188-207).

[15] A. DEAIBES Espaces uniformes et espaces de mesures, Publ. Dép. Math. Lyon 12-4 (1975) 1-166.

[16] J.W. DAVENPORT Multipliers on a Banach algebra with a bounded approximate identiy, Pac. J. Math. 63 (1976) 131-135.

[17] R.A. FONTENOT The double centralizer algebra as a linear space, Proc. Amer. Math. Soc. 53 (1975) 99-103.

[18] Topological measure thory for double centralizer algebras, Trans. Amer. Math. Soc. 220 (1976) 167-184.

[19] C.S. HERZ The spectral theory of bounded functions, Trans. Amer. Math. Soc. 94 (1960) 181-232.

[20] K.H. HOFMANN The duality of compact semigroups and C*-bialgebras, (Springer Lecture Notes 129, 1970).

[21] Categorical theoretical methods in topological algebra (Categorical topology - Springer Lecture Notes 540, 345-403).

[22] J.R. ISBELL Algebras of uniformly continuous functions, Ann. of Math. 68 (1958) 96-125.

[23] Uniform spaces (Providence, 1964).

[24] B.E. JOHNSON Centralizers on certain topological algebras, Jour. Lond. Math. Soc. 39 (1964) 603-614.

[25] An introduction to the theory of centralisers, Proc. Lond. Math. Soc. 14 (1964) 299-320.

[26] A. LAZAR and D.C. TAYLOR Double centralizers of Pedersen's ideal of a C*-algebra, I,II, Bull. Amer. Math. Soc. 78 (1972) 992-997 and 79 (1973) 361-366.

[27] S. MAC LANE Categories for the working mathematician, (Springer, 1971).

[28] S. MARDEŠIĆ Pre-categories and shape theory (in Springer Lecture Notes 540 (1976) 425-434)

[29] K. McKENNON The strict topology and the Cauchy structu of the spectrum of a C*-algebra, Gen. Top. and Appl. 5 (1975) 249-262.

[30] P. MICHOR Duality in groups (Unpublished note, 1972)

[31] A. NAZIEV The realization of dual categories, Soviet Math. Dokl. 14 (1973) 1492-1495.

[32] J.W. NEGREPONTIS (J.W. PELLETIER) Duality in analysis from the point of view of triples, Jour. Alg. 19 (1971) 228-253.

[33] G. PEDERSEN Applications of weak* semicontinuity in C*-algebra theory, Duke Math. Jour. 40 (1973) 431-450.

[34] R. PUPIER Methods fonctorielles en topologie général (Université de Lyon, 1971).

[35] M. ROME L'espace $M^\infty(T)$, Publ. Dép. Math. Lyon 9-1 (1972) 37-60.

36] H. SCHUBERT Kategorien I,II (Springer, 1970).

37] Z. SEMADENI Banach spaces of continuous functions I (Warsaw, 1971).

38] F.D. SENTILLES and D.C. TAYLOR Factorisation in Banach algebras and the general strict topology, Trans. Amer. Math. Soc. 142 (1969) 141-152.

39] D.C. TAYLOR The strict topology for double centralizer algebras, Trans. Amer. Math. Soc. 150 (1970) 633-643.

40] A general Phillips theorem for C^*-algebras and some applications, Pac. J. Math. 40 (1972) 477-488.

41] Interpolations in algebras of operator fields, Jour. Func. Anal. 10 (1972) 159-190.

42] J.L. TAYLOR Topological invariants of the maximal ideal space of a Banach algebra, Adv. in Math. 19 (1976) 149-206.

43] J.-K. WANG Multipliers of commutative Banach algebras, Pac. J. Math. 11 (1961) 1131-1149.

44] R.F. WHEELER The strict topology, separable measures and paracompactness, Pac. J. Math. 47 (1973) 287-302.

CONCLUSION

As mentioned in the introduction, we have attempted in this monograph to present a first synthesis of a theory of mixed topologies and their applications. We would like to conclude with a brief list of some general problems. Some of these topics are the object of current research.

We begin with the closed graph theorems and Banach-Steinhaus theorems of Chapter I. At present, there are essentially two approaches to such results - partitions of unity and some kind of geometric condition on the unit ball such as the Σ_1 condition (we take this opportunity to inform the reader that there is a Σ_2 condition which was introduced also by ORLICZ but which we have not treated here). It would be interesting to find some new ideas which lead to new closed graph theorems. In particular, neither of these approaches is applicable to spaces of holomorphic functions (in contrast to continuous or measurable functions). Hence the question: is there a closed graph theorem for abstract Saks spaces from which MOONEY's theorem (V.2.12) can be deduced ? As a second topic we have seen that almost none of the important classes of locally convex spaces (e.g. Fréchet, nuclear or barrelled spaces) have any relevance for Saks spaces. Now most of these classes can be classified within the category of locally convex spaces in a manner which can be formally carried over to the category of Saks spaces. It would be interesting to investigate the corresponding classes of Saks spaces.

We give some examples to indicate more precisely what is meant. The natural analogy of a Frêchet space would be a Saks space projective limit of a sequence of Banach space i.e. a complete Saks space $(E, \| \|, \tau)$ where τ is metrisable. We could also define a nuclear Saks space as a space for which various Saks space tensor products coincide (in analogy to the condition $E \hat{\otimes} F = E \hat{\hat{\otimes}} F$ for arbitrary F which characterises nuclearity for locally convex spaces). The first question would be to decide if this is a useful notion, the second to give an internal characterisation of these spaces. Of course, this definition depends on which tensor products for Saks spaces we consider to be relevant and in this connection we mention that in defining tensor products for Saks spaces in Chapter I we restricted attention to the two simple definitions which were tailor-made for the applications we had in mind - the tensor product representations of $C^\infty(S \times S_1)$ and $L^\infty(M \times M_1)$. It would perhaps be useful to have a more differentiated treatment, with particular regard to universal properties for example.

With regard to Chapter II, we list, in a rather haphazard way, some problems which seem to us to be relevant. A fruitful branch of modern point-set topology has been the study of coverings of a topological space. We can define strict topologies on $C^\infty(S)$ which are intimately related with covering properties of S as follows: if $\mathcal{U}$ is (for example, an open) covering of a completely regular space S, we can define a Saks space structure on $C^\infty(S)$ by mixing the supremum norm with the topology of uniform convergence on the sets of $\mathcal{U}$. Given a suitable family of coverings, we can define a Saks space structure on $C^{\infty}(S)$ by taking the locally convex inductive limit of the structures induced by the individual coverings (possible candidates for the distinguished family of coverings could be all open coverings or all locally finite open coverings). The problem would be to establish relationship between the topology of S and the properties of $C^{\infty}(S)$ under the corresponding strict topologies and to characterise the duals as spaces of measures on S, in particular to establish conditions for the equality of these strict topologies with those introduced in Chapter II. We conclude our remarks on Chapter II with two problems: firstly to develop an approach to cylindrical measures on a locally convex space E using a strict topology on $C^{\infty}(E)$ and secondly to give a detailed analysis of the space of continuous linear operators from $C^{\infty}(S)$ into a Banach space E and its relation with vector measures, corresponding to the theory for compact S. J.L. TAYLOR has recently shown how topological properties of a compact space are reflected in the properties of a Banach algebra which has this space as spectrum - in particular, the cohomology and K-theory of the space. Can this theory be carried over to non-compact spaces using strict topologies ?

The outstanding question raised in Chapter III is that of developing a complete theory of integrable and measurable functions with valued in a Saks space, thus obtaining a synthesis of various notions of vector-valued integration. It would also be useful to make more precise the relation with vector measures, in particular, with regard to the Radon-Nikodym property.

With regard to Chapter IV it would be particularly interesting to try out the various Saks space tensor products on W*-algebras and to compare the results with the tensor products of W*-algebras which have been studied in the literature. Another interesting question is that of studying the corresponding mixed topologies on spaces L(E) or even L(E,F) (E and F Banach spaces) to see how much of the theory can be carried over. For example, can the dual of L(E) or L(E,F) be identified with a space of operators (the approximation property will presumably play a role here)?

As regards Chapter V, we remark that in the two cases ($C^{\infty}(S)$ and $H^{\infty}(U)$) where we have investigated the spectrum of a commutative Saks algebra, we have found that $M_{\gamma}(A)$ was dense in the Banach algebra spectrum M(A). In the first case the result is not very deep (it simply means that S is dense in βS). The second case is of course the Corona problem. This leads us to pose the following problem: is $M_{\gamma}(A)$ always dense in M(A)? It may be that there is a simple counter-example. Of course, if the result is true it must be very deep and presumably can only be solved when we have considerable information on the structure of Saks algebras. Another problem is to study the spectrum of $H^{\infty}(G)$ where G is a domain in higher dimensions or even a more complicated domain in $\mathbb{C}$. In general, $M_{\gamma}(H^{\infty}(G))$ is much larger than G e.g. if there is a domain G_1 properly containing G with the property that each bounded holomorphic function on G has a bounded holomorphic extension to G_1. We thus pose the following problem: can $M_\gamma(H^\infty(G))$ be regarded as some kind of bounded envelope of holomorphy of G; more precisely, does $M_\gamma(H^\infty(G))$ have a complex analytic structure so that the Gelfand-Naimark transform of elements of $H^\infty(G)$ are holomorphic functions on $M_\gamma(H^\infty(G))$? More generally, we can ask for general conditions on a commutative Saks algebra which ensure that its spectrum has such an analytic structure. Another problem which is related indirectly to the theme of Chapter V is that of applications of mixed topologies to spaces of smooth functions and distributions. As an example, we could consider the space of bounded C^∞-functions on $\mathbb{R}^n$ (or on a manifold) and mix the supremum norm with the topology of uniform convergence on compacta of all derivatives. The dual is a space of distributions on the manifold. However, there is one problem - the auxiliary topology is not coarser than the norm topology so that the conditions of Definition I.1.4 are not satisfied. However, many of the results of Chapter I do not require this condition or can be modified in an obvious way and, in fact, a number of the references given in Chapter I contain results on this more general situation. Just as in measure theory where the mixed topologies are useful in the generalisation of the theory of Radon measures from locally compact to completely regular spaces, so one would expect that such mixed topologies could be helpful in the study of distributions on a Banach space (or on a manifold modelled on a Banach space) where the classical Schwartz approach breaks down because of the absence of non-trivial test functions on a Banach space (analogous to the non-existence, in general, of continuous functions with compact support on a non-locally compact space).

It has been our intention in this monograph to expose the main outlines of the theory of Saks spaces and some of their applications. We hope that we have succeeded in demonstrating that it is the natural tool for some problems which are rather intractable when treated by classical methods. The fact that we have not included deeper applications to various problems on function spaces is, we believe, rather a testimony to the inadequacy of the author than a condemnation of the potential of the theory. We hope that this monograph will induce some experts to explore the possibility of applications of Saks spaces to such problems.

INDEX OF NOTATION

INDEX OF TERMS

NORTH-HOLLAND MATHEMATICS STUDIES

A series of paperback tutorials in pure and applied mathematics

Vol. 16: **Volterra Stieltjes - Integral Equations**
by C.S. HÖNIG
Notas de Matemática [56] - 1975 x + 158 pages

Vol. 17: **Hewitt-Nachbin Spaces**
by M.D. WEIR
Notas de Matemática [57] - 1975 viii + 270 pages

Vol. 18: **Global Theory of a Second Order Linear Ordinary Differential Equation with a Polynomial Coefficient**
by Y. SIBUYA
1975 xvi + 290 pages 63 illus. 50 lit. refs.

Vol. 19: **Intensional and Higher-Order Modal Logic**
With Applications to Montague Semantics
by D. GALLIN
1975 xi + 148 pages

Vol. 20: **Vector Measures and Control Systems**
by I. KLUVÁNEK and G. KNOWLES
Notas de Matemática [58] - 1976 x + 180 pages

Vol. 21: **New Developments in Differential Equations**
Proceedings of the Second Scheveningen Conference on Differential Equations, The Netherlands, August 25-29, 1975
edited by W. ECKHAUS
1976 viii + 248 pages 6 illus. 157 lit. refs.

Vol. 22: **Hopf Spaces**
by A. ZABRODSKY
Notas de Matemática [59] - 1976 x + 222 pages

Vol. 23: **Near-Rings**
The Theory and Its Applications
by G. PILZ
1977 xiv + 394 pages over 700 lit. refs.

Vol. 24: **Topological Algebras**
by E. BECKENSTEIN, L. NARICI, C. SUFFEL
Notas de Matemática [60] - 1977 xii + 370 pages

Vol. 25: **Approximation of Vector Valued Functions**
by J.B. PROLLA
Notas de Matemática [61] - 1977 xiii + 219 pages

Vol. 26: **Bornologies and Functional Analysis**
by H. HOGBE-NLEND
Notas de Matemática [62) - 1977 xii + 144 pages

Vol. 27: **Functional Analysis: Surveys and Recent Results**
edited by K.-D. BIERSTEDT *and* B. FUCHSSTEINER
Notas de Matemática [63] - 1977 xii + 290 pages

ISBN 0 444 85100 3